Environmental Finance for the Developing World

Environmental Finance for the Developing World

Michael Curley

CRC Press is an imprint of the
Taylor & Francis Group, an **informa** business

CRC Press
Taylor & Francis Group
6000 Broken Sound Parkway NW, Suite 300
Boca Raton, FL 33487-2742

CRC Press is an imprint of Taylor & Francis Group, an Informa business

Printed on acid-free paper

International Standard Book Number-13: 978-0-367-33686-8 (Hardback)

Library of Congress Cataloging-in-Publication Data

Visit the Taylor & Francis Web site at
www.taylorandfrancis.com

and the CRC Press Web site at
www.crcpress.com

Contents

Foreword

Water is a prerequisite for human survival and dignity and a fundamental basis for the resilience of both societies and the environment. Water is vital for human nutrition and health, and essential for ecosystem management, agriculture, energy, and overall planetary security. Ensuring sustainable access to safe water and sanitation, achieving sustainable water management, and preventing pollution, scarcity and reducing flooding events are key global challenges of the 21st century.

Currently, more than two billion people lack access to safe drinking water and more than double that number lack access to safe sanitation. With the expected urban expansion to increase from 55% to 68%, urban planning will need to integrate system-wide water management approaches to limit the footprint that cities have on nearby water quality, quantity, and on energy and agri-food systems. The spatial planning of land development will be critical for protecting water resources. Investments and financing in nature-based solution and green infrastructure, forested areas and wetlands, are important in capturing runoff into water supplies and preserving ecosystem services to enhance natural capital, reduce costs, and support a resource-efficient and competitive circular economy.

These interwoven benefits, which are the essence of sustainable development, are central to achieving Agenda 2030.

Harnessing water to drive sustainable growth and delivering universal access to services requires investments in water infrastructure, information, and institutions. Yet current levels of water-related investment fall short of delivering the global water agenda, in particular, Sustainable Development Goal (SDGs) 6 on water. Global estimates for financing needs for a broader range of water infrastructure span from $6.7 trillion by 2030 to $22.6 trillion by 2050. Investment is required to protect populations, cities, economies, and ecosystems from water-related risks – risks of droughts, floods, pollution, inadequate access to safe drinking water supply and sanitation, and degradation of water-dependent ecosystems.

Strengthening the enabling environment for investment is needed in order to provide incentives for the sustainable management of water resources, minimizing overall investment needs and reducing the risk of investments failing to deliver expected benefits.

Hence this book is designed to facilitate the understanding of basic financing principals to aid project and program administrators in developing and maintaining investment funds for water projects, building on the processes and procedures in infrastructure financing and in developing sustainable grants and loans to service providers for viable water-related projects. The scaling up of such efforts through wide dissemination of financing mechanisms and principles offered in this technical assistance handbook is useful for any academic background administrator, whether a civil engineer, a legal counsel, an environmental and social specialist, government fonctionnaire, and/or an economist.

The economic transformation to improve water resource management and to stimulate private sector investment all require the basic understanding of project financing elements covered in this handbook: planning a sustainable and targeted water and sanitation fund; ensuring proper management and implementation of the fund; scoping stakeholder and partner concerns and interests; and developing recommendations in leveraging and de-risking operationalization and investment for growth in the sector. This book provides the foundation for understanding the broad range of financing instruments and blended public and private capital to generate catalytic effect. Responsible financing and ensuring an alignment with development goals in leveraging finance and de-risking water investments for achieving the SDGs must be coupled with capacity building, for which this book intends and provides the material support for such undertaking.

Sasha Koo-Oshima
Head of Water and Deputy Director of
the UN Food and Agriculture
Organization
(former U.S. EPA Office of Water, Senior Advisor in establishing State Revolving Water Funds in international water financing assistance)

Preface

This book is about creating a decent living environment for the hundreds of millions of people who live on this planet without safe drinking water or basic sanitation services. The technology for creating safe drinking water and basic sanitation systems isn't exactly child's play, but it's not rocket science either. The technology isn't the problem: money is the issue.

During my career I have had the good fortune of either heading up, or being part of, teams that provided safe drinking water to rural villages in developing countries. Fifteen were in Kazakhstan and several more, as a board member of the former International Rural Water Association, were in Nicaragua, El Salvador, and Guatemala.

In Kazakhstan, it was a government-to-government program. The US and the European Union had become concerned that with the fall of Communism, the mechanism for financing basic water and sanitation systems had collapsed as well. They wanted to see new systems created that could be integrated with the international capital markets – not just some old socialist financial relic. I always think that the EU and the US must have flipped a coin and that the US lost. So, they got the job. The US State Department then turned the matter over to EPA.

At that time I was on the Environmental Financial Advisory Board (EFAB) at EPA, where, despite the fact that appointees were limited to two three-year terms, I served for 21 years under four Presidents. I had also just published my first book, the *Handbook of Project Finance for Water and Wastewater Systems*. The Office of International Affairs at EPA – on whom the burden fell – heard about me from the EFAB folks and conscribed me for the job of creating new environmental infrastructure finance programs in the former Soviet Union.

In Kazakhstan, the central government took care of the environmental infrastructure in the major cities of Almaty, Astana, and Atyrau. But the rural areas were left to fend for themselves. So, the US/EU entered into an agreement with the Kazakh government to set up a rural water finance program with demonstration projects in 15 villages in the Almaty Oblast (region). The money for the demonstration projects would come 60% from the US Government, 20% from the Kazakh Government, and 20% from the Oblast Government.

As noted above, the technology was no problem in Kazakhstan. All you had to do was pay for it. So, the projects got built. But what then? What happens when a pipe breaks, or the system runs out of chlorine? In our case, we specifically chose villages where the village council would agree in advance to set up a modest fee collection system from all village residents to pay for the chemicals, the electric power, and the maintenance on their systems. And once the money was there, the know-how was no problem.

So, the program worked in Kazakhstan. And it's still working.

In Central America with the IRWA, the situation was different.

The IRWA was the daughter corporation of the National Rural Water Association (NRWA), which represents thousands of small water systems across the US (There are over 50,000 water systems in the US) The NRWA exists on its modest dues income as well as grants from EPA and the US. Department of Agriculture. In those days, the NRWA furnished the IRWA with a modest budget to do water infrastructure projects in Central America.

We had no problems with technology in Central America since we brought it with us. As you might imagine, here in the US, with small, rural water systems, the CEO is often a jack-of-all-trades. And so it was that some special members of the NRWA donated their time to go to Central America to build rural water systems in poor villages there. So, we, at the IRWA, had both the money and the expertise to build systems in Central America.

We also made sure that every village we worked in had some kind of a "water committee" that had the full support of the citizens there – just like in Kazakhstan. But then we ran into a problem. It wasn't money this time; it was technology. No one in our villages had a clue what to do to maintain their system or to fix it, if there were a problem.

Now, the NRWA created – and is famous for – a wonderful institution: its "Circuit Rider" program. Circuit Riders are technicians who know everything about managing, maintaining, and repairing/replacing rural water systems. We had the pleasure of working with one of the NRWA's great Circuit Riders whose "circuit" was 23 counties in northeast New Mexico. He went around to all of the small water systems in those 23 counties, fixed their problems, and otherwise made sure they were functioning properly.

There were no local Circuit Riders in Central America. What's more, the name "circuit rider" doesn't translate into Spanish (or any other language) at all! The term "circuit rider" comes from the 19th century days in the old American West. Circuit riders were either lawmen or judges who worked in a large territory, and who organized their assigned work areas into "circuits" around which they rode their horses to get their jobs done.

As the IRWA program was coming to an end because of money problems at the NRWA, we were working on the establishment of a training center for village water technicians in Central America. Since the words "circuit rider" have no meaning in Spanish, we couldn't call them that. Furthermore, they were never going to be high-paying jobs. So, the only reason for doing this work would be the pride of providing people with an essential service. We, therefore, were going to give these technicians the honorific titles "Caballeros de Agua", which literally means "Knights of Water". Alas, it never came to pass.

As noted above, the technology can always be found to build safe drinking water and basic sanitation systems – in any part of the world. So, it's not the technology, it's the money that's the problem. And therefore, it is absolutely critical to do all that is humanly possible to make such environmental infrastructure as cost effective as possible. That is the purpose of this book. There is an axiom in environmental finance: the less expensive the projects, the more projects will get done. The more projects that get done the better the quality of life for the people on this fragile planet. That said, let me conclude with a short, personal story about the quality of life on this fragile planet:

In '03 we were starting to build village water systems in rural Kazakhstan. There were lots of logistical problems to clutter my head. When the first one was done, I went over to inspect it. My team said I had to wear a jacket and tie. I was the big dude from New York.

After looking everything over, the Village Mayor invited me and my team leader to lunch at his house. His wife was the village medical officer. After lunch she took me down to her clinic and showed me the log of medical cases she kept. Page after page of diarrhea and dysentery cases. Then she pointed out the date when our new water system went operational. From then the log's pages were almost blank. The diarrhea and dysentery went from 5–10 a day to 1–2 a MONTH!

Only then did I really begin to realize what we had done for the 2500+ poor souls in the village.

Then, on the way back to the truck, this old grandmother came up to me, kneeled down in the mud, and kissed my hand, muttering "thank you, thank you" in Russian.

And so, I commend the following pages to you. I only hope that you enjoy reading them as much as I enjoyed writing them.

Acknowledgments

The one person who is most responsible for this book is my great friend and colleague of many happy memory, Bill Freeman, who was in charge of the Former Soviet Union program at the turn of the century at the Office of International Affairs at the US Environmental Protection Agency (EPA). In the Fall of 2007 Bill went into hospital for some minor back surgery. The following day he was home, sounding good and feeling fine. He died that afternoon of a heart attack.

As you will see, Bill recruited me to work in the Former Soviet Union reorganizing their procedures for financing environmental infrastructure. He worked with colleagues in the Organization for Economic Cooperation and Development (OECD) in Paris. OECD was formed in 1961 as an outgrowth of the North Atlantic Treaty Organization (NATO) to foster economic development and world trade for all of Europe. Based on my work for them, I was asked to assemble the materials I had developed for the various countries into a three to five day course on the fundamental financial principles required to build environmental infrastructure anywhere in the world. That is what you are now reading.

In addition to working in the Former Soviet Union, I had the honor of serving on the board of the International Rural Water Association (IRWA). Bill Kramer was the Executive Director of that organization. But he was far more than just a manager. He was a true leader and the source of much inspiration and sound thinking for that group.

Randy VanDyke was my great colleague on the IRWA board. Charles Hilton, L.C. "Buddy" Hand, and Doug Anderton were also valuable colleagues on the IRWA board.

I also had the great pleasure of working with Steve Gustafson, head of the water program at the National Bank for Cooperatives (CoBank) and Jim Maras, who was the head of the rural water program at the US Department of Agriculture, and then went on to run the water program at CoBank as well. I still have the pleasure of working with Randy VanDyke and Jim Maras today on other important matters.

Finally, I would be truly remiss without mentioning Lupe Aragon and the late Fred Stottlemyer. Fred was a member of the National Rural Water Association (NRWA), which was the parent organization of the IRWA. Fred ran a rural water system in West Virginia. Fred knew everything about building and running water systems. He volunteered for the IRWA to build water systems in poor villages in Central America.

Lupe Aragon was an employee of the NRWA. His title was "Circuit Rider", which you will learn about. Lupe's job was to keep all of the rural water systems running in 23 counties in northern New Mexico. He also volunteered to work for the IRWA in poor communities in Central America.

I have special gratitude for Lupe and Fred. Some years ago, one of the young women at our church returned from a Summer trip to an orphanage in Nicaragua complaining that all of the children were getting sick there every day from drinking water out of the taps. Her father sent her to me. I contacted Fred and Lupe. The three of us went down to the orphanage. Fred and Lupe diagnosed the problem and said the fix would cost $10,000. My church raised the money. Fred and Lupe and I went back down to Nicaragua to supervise the construction. Not a single child has gotten sick from the water there since then. Furthermore, the orphanage used to try in vain to get the children to drink bottled water and spent a lot of money on it. After Fred and Lupe fixed the problem, the orphanage didn't have to spend over $400 per month on bottled water. Instead they used the money to pay for a doctor to come in once a week to care for the kids.

Introduction

The concept of this book began with the idea of teaching people how to get the money together to build basic, safe drinking water and sanitation systems in poor villages throughout the developing world.

There is a widespread misconception that people – especially in small, poor villages in developing countries – cannot afford to pay for the essential human services they need to survive. In almost all circumstances, this isn't true. They can. They don't have to walk miles each day to get drinking water. They don't have to relieve themselves in their gardens or back yards or other unsanitary places.

The problem is to design the most cost effective project. This means, first, that the project must deliver only the basics. No frills. Second, the project needs to be as inexpensive as possible. Third, the project must be paid for with borrowed funds, where the underlying loans have the lowest possible interest rates and the longest feasible terms.

This book is about the third factor: the most favorable financing.

(Borrowed funds? Some readers might be asking: what about grants? Grants are free. They don't have to be repaid. They are obviously the cheapest source of money. True. But grant funds are rare. They depend on largesse and willing donors who have excess cash. They also depend on a host of social and political considerations. That is why they are so difficult to obtain. If people in small, poor, rural villages in developing countries are intent on waiting for grants, they will probably die waiting – and they may well die from their unsafe drinking water or unsanitary conditions.)

There are about 800 million people without access to safe drinking water and over 1.8 billion people without basic sanitation services. They aren't all going to get grants.

There are ten concepts involved in the financing of these basic human services. Hence there are ten chapters in this book, plus a first chapter that sets the stage for the others. The book is designed to teach these subjects. Each chapter is a self-contained teaching unit. You will note that each chapter explains what material it will present then it goes on to relate the contents of that chapter to the contents of the remainder of the book. Then it presents the specific materials designed for that specific chapter.

This book is designed for three audiences. The beginning of each chapter will specify which audience will find that chapter's material most useful.

The first audience consists of local government officials who work for water and wastewater utilities and who are involved in financial project preparation.

The second audience consists of the senior executives or senior local government officials who are responsible for managing water and wastewater utilities. For this audience, almost all of the chapters should provide valuable background information for their positions.

The third audience consists of government executives and representatives at the national, regional, and even local levels, who are responsible for making policy for environmental infrastructure.

1 Financing the Global Environment

INTRODUCTION

This chapter is the first in a series of 11 teaching units on the financing of environmental infrastructure projects. Each chapter contains all the elements of a teaching. They are:

1 Introduction to Environmental Project Finance
2 Measuring Income
3 Maximizing Cash Available for Debt Service
4 Loan Basics
5 Project Valuation
6 Financial Feasibility
7 Alternate Finance Options
8 Sources of Funds

Module Audience

Three distinct audiences of the financial modules:

- Executives and local governments working for water and wastewater utilities concerned with financial project preparation and financial analysts working at government funding agencies or programs (Modules I – VI)
- Government policy makers, especially those who make policy concerning water and wastewater utilities and their relationships with their users or customers (Modules VII – XI)
- Senior executives or senior local government officials responsible for managing water and wastewater utilities (All Modules provide valuable background information)

9 Cost/Benefit Analysis
10 Tariff Design
11 Subsidies

This book has been written with three distinct audiences in mind.

The first audience consists of executives or local government officials who work for water and wastewater utilities and who are concerned with financial project preparation. For this audience, Chapters 1 through 6 should be useful. Financial analysts at government funding agencies or programs will also find them helpful.

The second audience consists of government policy makers, especially those who make policy concerning water and wastewater utilities and their relationships with their users or customers. For this audience, Chapters 7 through 11 should be helpful.

The third audience consists of the senior executives or senior local government officials who are responsible for managing water and wastewater utilities. For this audience, all of the chapters, with the possible exception of Chapter 8 – Sources of Funds, should provide valuable background information for their positions.

Module Overview

This module is the first of eleven covering the following topics:

- **Module I** – *Introduction to Environmental Project Finance*: Introduces the concept of project finance
- **Module II** – *Measuring Income*:
  Defines what income can be used for making debt service payments and how it can be estimated
- **Module III** – *Maximizing Cash Available for Debt Service*:
  Stresses the importance of maximizing income to allow utility systems to incur debt to finance needed projects
- **Module IV** – *Loan Basics*:
  Devoted to the basic concepts of loans and debt, including an explanation of the present value theory of money

2

The first six chapters in this book constitute the first six steps in developing the financial information necessary for undertaking projects where income is earned from the delivery of certain utility services, whether the utility is publicly or privately owned. The focus of these training chapters will be on water and wastewater utilities; but the same principles apply to other similar types of utilities such as those that deliver power, heat, or solid waste services. Those six chapters are written from the perspective of a utility executive who must undertake a project to improve this system and is interested in learning the financial techniques necessary to obtain the money to undertake the project.

After introducing the concept of project finance (Chapter 1), the first step is the presentation of methods for measuring income (Chapter 2 – Measuring Income). All money is not the same. Not all income can be used for project debt. Chapter 2 defines what income can be used for making debt service payments and how it can be estimated.

Methods for maximizing income are presented in the succeeding chapter (Chapter 3 – Maximizing Cash Available for Debt Service). It is emphasized that the maximization of income is of paramount importance because if income exceeds operating expenses, needed projects can be undertaken to improve utility systems by having the utility incur debt. If utilities have excess income, they can be free of the pernicious reliance on grants. Excess income can be used to pay annual debt service. The next chapter, therefore, is devoted to the explanation of the basic concepts of loans and debt (Chapter 4 – Loan Basics). The present value theory of money is also explained in that chapter.

Module Overview

- **Module V** – *Project Valuation*:
  Evaluates the various types of loans in order to determine which available loan is best for a particular utility
- **Module VI** – *Financial Feasibility*:
  Uses all of the information presented in the preceding five modules to illustrate which projects are financially feasible for a utility to undertake
- **Module VII** – *Alternate Finance Options*:
  Introduces the basic concepts of grants, subsidized interest rate loans, market rate loans, and loan guaranties
- **Module VIII** – *Sources of Funds*:
  Customary sources of funding for environmental projects is listed and discussed
- **Module IX** – *Cost/Benefit Analysis*:
  Presents basic rubric of how to optimize meager funds available for environmental projects and prioritize them using matrices

3

Once the basic concepts of borrowing are understood, the next step is to be able to evaluate various types of loans in order to determine which available loan is best for a particular utility. Techniques for evaluating loans, therefore, are presented in the following chapter (Chapter 5 – Project Valuation).

Chapter 6 – Financial Feasibility, uses all of the information presented in the preceding five chapters to illustrate which projects are financially feasible for a utility to undertake. It demonstrates both how the largest possible project could be undertaken at the lowest possible cost and how a utility's excess operating income could be used to improve the system.

In Chapter 7 – Alternate Finance Sources, the perspective shifts. Financial concepts are no longer presented from the utility executive's perspective. Rather, in this chapter, financial concepts are presented from the perspective of a government official whose responsibility it is to manage or administer a program to provide funds for municipal utility services. In other words, in the first six chapters, project finance is discussed from the perspective of one needing funds (for a particular project). In the seventh chapter, the concepts of project finance are discussed from the perspective of a government official who has funds, or whose job it is to seek funds for utility projects in his country. It is this official's responsibility to provide the best quality of utility services to the largest number of people by providing funds for the utilities themselves, so that they, in turn, can improve services.

Chapter 7 contains some of the most important information in this entire training series. The basic concepts of grants, subsidized interest rate loans (concessionary loans), market rate loans, and loan guaranties are all presented. Each alternative is then compared to all of the others to determine which funding alternative is most efficient, i.e. which funding method buys the largest project for the smallest amount of money.

In the eighth chapter, Sources of Funds, the same perspective is maintained as in the previous chapter. In this chapter, the customary sources of funding for environmental projects are listed and discussed. The major thrust of the chapter is the significant role that the private sector can play.

Chapter 9 – Cost/Benefit Analyses is, perhaps, the most important element of this 11-module series. In this chapter, the government official audience is presented with a basic rubric of how to optimize often-meager funds available for environmental projects. Matrices are devised to prioritize the "projects which bring the greatest amount of public health benefits to the largest number of people for the least amount of grant funds".

Module Overview

- **Module X** – *Tariff Design*:
  Introduces the methodology of creating fair, volumetric tariffs
- **Module XI** – *Subsidies*:
  Describes the role of subsidies in maximizing utility revenues

4

In Chapter 10 – Tariff Design, we return to addressing the needs of local government officials and utility managers. In Chapter 3 – Maximizing Cash Available for Debt Service, the need for creating fair, volumetric tariffs that provided enough income to cover the cost of operating and maintaining the utility system was discussed. In Chapter 10, the methodology of creating such tariffs is presented. There is also a discussion of how such tariffs can be regulated by local governments to assure users of the basic fairness of the tariff design.

In Chapter 11 – Subsidies, another matter that was covered briefly in Chapter 3 – Maximizing Cash Available for Debt Service, will be discussed: the role of subsidies in maximizing utility revenues.

The conceptual basis for any subsidy is that the goods or services subsidized are otherwise unaffordable to consumers.

The thesis of this chapter is that only those truly in need should be subsidized. Those who can afford the service should pay for it. The chapter discusses how to design subsidies that target only the truly needy.

2 Measuring Income

INTRODUCTION

In this chapter we will see how to determine how much cash is available to pay the debt service – principle and interest – of money borrowed to fund a project. In order to determine how much funding is needed for a project it is necessary to categorize the existing incomes and expenses. This will allow you to determine how much of your current income can be used to pay for debt service. **Cash Available for Debt Service** is a specific term among credit analysts that refers to the funds used to pay for a new project. This concept is critical to all other project funding decisions.

CHAPTER CONTENTS

- Total Income
- Regular Income
- Non-Recurring Income
- Total Expenses
- Non-Cash Expenses

- Cash Expenses
- Cash Available for Debt Service
- Accounting for Bartering

EXAMPLE: CALCULATING CASH AVAILABLE FOR DEBT SERVICE

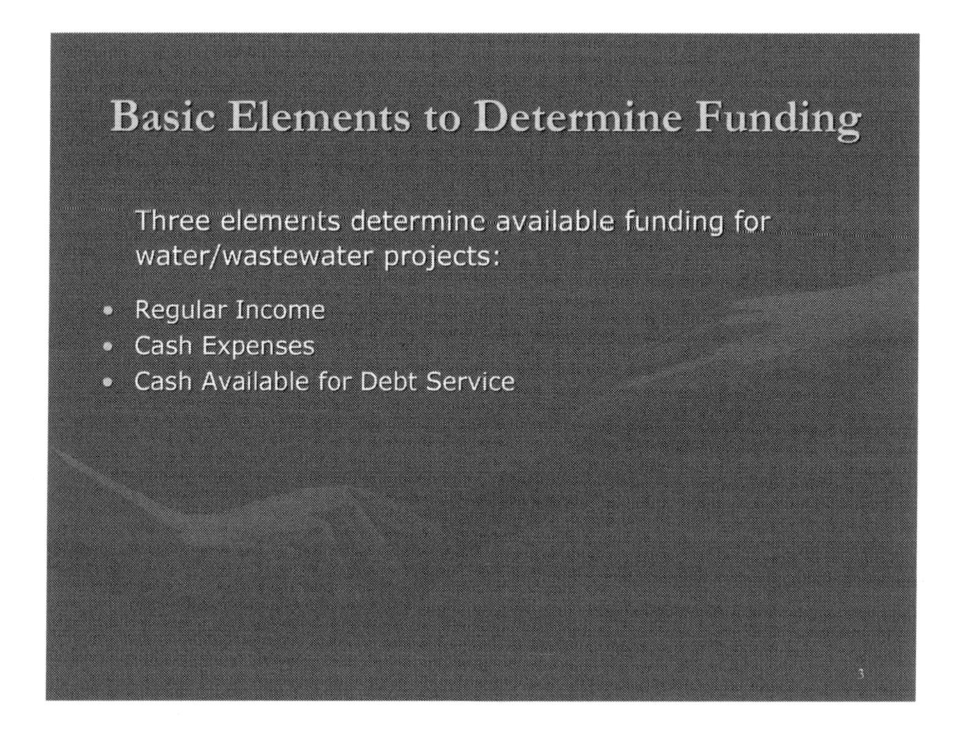

In order to determine what funds are available to pay for a water or wastewater project, it is important to understand three elements:

- Regular Income
- Cash Expenses
- Cash Available for Debt Service

Regular Income

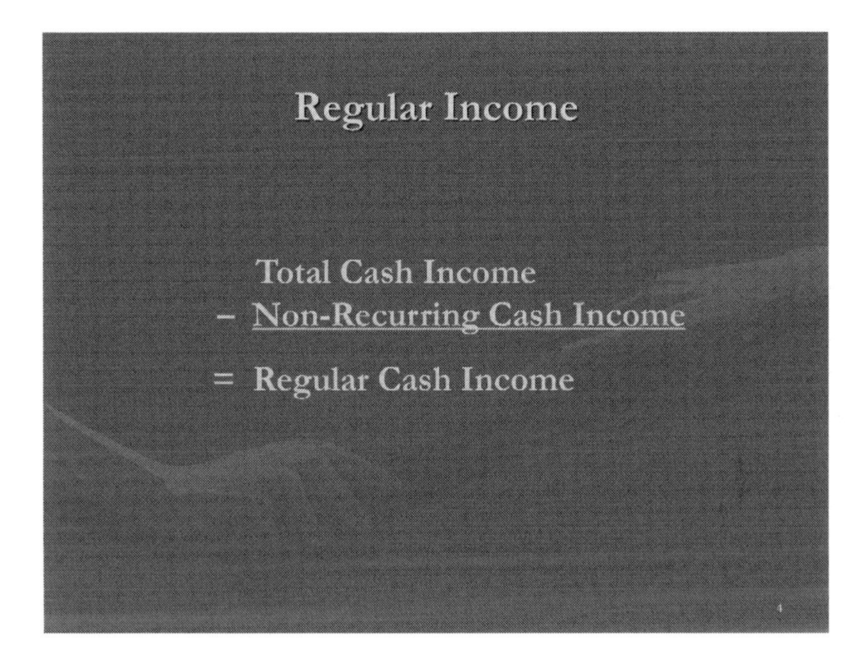

The first step to calculating Cash Available for Debt Service is identifying Regular Cash Income. Regular Cash Income encompasses all monies that are collected from the daily operations and management of the utility. The term regular does not mean ordinary but instead means predictable and continuous. Tariffs and interest that have proven to be constant and reliable sources of income can be projected Regular Cash Income in the future. Non-Recurring Cash Income cannot be used to predict future income and therefore cannot be counted on as cash available for debt service. Non-Recurring Cash Income includes such things as the income from the sale of assets such as land or equipment, government grants, and user fees. Only in short term cases of very constant growth, can connection fees and similar fees for service be considered regular income.

Total Cash Income

- All monies received by the utility
- Includes:
  - *Tariffs*: Income from providing service (water) to ratepayers
  - *Interest*: If utility has funds in bank accounts or other investments, then the interest from those accounts is income
  - *Fees for Service*: Fees collected for services not related to the tariff such as connection fees
  - *Sale of Assets*: Money from selling old equipment, land, or other assets
  - *Miscellaneous*: All other monies collected by the utility

5

Total Cash Income includes:

- *Tariffs*: The per unit income of providing service to ratepayers (see Chapter 10 for discussion of tariff design).
- *Fees for service*: The fees collected for services not related to the per unit tariffs, such as connection fees.
- *Interest*: If the utility has funds invested or saved, then the interest from those accounts is income.
- *Sale of assets*: The profits from selling old equipment, land, or other assets.
- *Misc*: All other monies collected by the utility.

Regular Cash Income = Total Cash Income – Non-Recurring Cash Income

Cash Expenses

Total Cash Expenses are ALL the expenses that a utility must pay, including the acquisition of assets, maintenance of machinery, electricity, chemicals, depreciation and amortization, and income to employees.[1]

Non-Cash Expenses are those expenses that do not result in cash expenditures such as depreciation and amortization.

1 It is important to recognize that prices change and future expenses should increase in a predictable manner.

Only Cash Expenses are used to determine Cash Available for Debt Service.

Cash Expenses = Total Expenses – Non-Cash Expenses

Cash Available for Debt Service

Regular Cash Income

- The income that arises or will arise from the day-to-day delivery of utility services
- Includes:
  - All *predictable* and *continuous* monies collected through tariffs from the daily operations and management of the utility, i.e., delivering water to ratepayers
  - Interest and other monies received by the utility that have proven to be constant and reliable sources

6

The last step is subtracting Cash Expenses from Regular Income to arrive at the Cash Available for Debt Service.

Non-Recurring Cash Income

- The income that is not regular or is otherwise not contingent on an event
- Cannot be counted as CADS or used to predict future income
- Includes:
  - Income from the sale of assets such as land or equipment, government grants, etc.
  - Connection fees and similar fees for specific services

7

Cash Available for Debt Service = Regular Income – Total Cash Expenses

To summarize, these terms have very specific meanings in the field of finance:

- *Total Cash Income*: All monies which are added to the accounts of the utility.
- *Non-Recurring Cash Income*: The income that is not regular or is otherwise contingent on an event.
- *Regular Cash Income*: The income that arises or will arise from the day-to-day delivery of utility services.
- *Total Cash Expenses*: Expenses which must be paid; does not include "*Non-Cash Expenses*" such as depreciation or amortization.

Cash Expenses

Total Expenses
– Non-Cash Expenses
= Cash Expenses

8

Non-Cash Expenses

- Expenses that do not result in cash expenditures such as depreciation and amortization
- Non-Cash Expenses do not determine CADS

10

- *Non-Recurring Cash Expenses*: These are part of Total Cash Expenses but not associated with day-to-day operations.
- *Cash Available for Debt Services*: Defined as Regular Income minus Total Cash Expenses, these funds are the **only** funds that can be used for calculating project payments and costs.

Total Cash Expenses

- All the expenses which must be paid in cash. This does not include "Non-Cash Expenses" such as depreciation or amortization.
- Includes cash paid out for:
  - Purchases of equipment
  - Maintenance of machinery
  - Electricity
  - Chemicals
  - Salary and wages of employees
- Only cash expenses are used to determine Cash Available for Debt Service

9

ACCOUNTING FOR BARTERING

Accounting for Bartering

- A ratepayer may choose to pay for utility with goods or services instead of cash
- These goods and services can either be *tangible* or *intangible*

12

Accounting for Bartering

There are two ways for the utility to account bartered items:

- If a utility receives goods, it can sell them directly and note the income as cash income
- Or, the utility can offset other expenses with the goods and note the decrease in total expenses

13

Sometimes income is in the goods or services instead of cash. A ratepayer may choose to pay for water with goods or a vendor may negotiate an arrangement where water is exchanged for a service such as electricity or chemicals. When this happens their "payment" can be measured by offset expenses. These goods are either *tangible*, physical items, or *intangible*, incorporeal goods or services. Regardless of the type, the utility accounts bartered items in one of two ways:

1. The utility can sell the goods directly and note the *profits* as cash income; or
2. The utility can offset other expenses with the goods and note the decrease in total expenses.

Sell the Goods

If bartered goods are easily sold, then the profits resulting from the sale of the items can be accounted as cash income.[2] It is important that one does not count revenue from the sale, but deduct the costs of the sale from the revenue to calculate the profits.

$$\text{Profits} = \text{Revenue} - \text{Costs}$$

Sale of Goods

- Money resulting from the sale of the items can be accounted as cash income
- Deduct the costs of the sale from the revenue to calculate the profits:

Cash Income = Money Received – Costs

14

Suppose the utility barters off a piece of land for several crates of wheat and then sells it for (USD) 500. The (USD) 500 is the revenue from the sale.

2 Depending on whether the goods or services are regular and predictable or a one time occurrence the income can be treated as Regular or Non-Recurring Income.

The utility then advertises the wheat for sale and an employee spends an afternoon finding a buyer and arranging the sale. This costs the utility (USD) 50 in income for the employee and (USD) 10 for the ad. The profit for the sale of the wheat is (USD) 500–60 or (USD) 440. Thus (USD) 440 is considered cash income from the sale of the land.

Offset Other Expenses

Offset Other Expenses

- A utility can use the bartered goods to offset other expenses that the utility would have otherwise had to pay in cash
- Barters commonly occur between the utility and another producer which require each other's services
- For example: the water utility may barter with the electric utility
- When goods that individuals, not utilities, need and use are bartered, they can be given to willing employees in *lieu of a portion of their salary or wages*

15

Rather than simply sell the goods, a utility can use them to offset other expenses that the utility would have otherwise had to pay in cash. Barters commonly occur between the utility and another producer, which require each other's services. For example, the water utility may choose to barter with the electric utility to trade water for electricity. For accounting purposes, the water utility manager deducts the amount that the utility would otherwise have to pay for electricity from its cash expenses.

When goods that individuals, not utilities, need and use are bartered, then they can be given to willing employees *in lieu of a portion of their cash income.* For an income example, a ratepayer pays a connection fee with four loaves of bread. The responsible official gives those four loaves of bread to a willing employee who would take the bread in lieu of some income, perhaps (USD) 10. The (USD) 10 that the utility didn't have to pay the employee can be considered a reduction in expenses from the connection fee if, and only if, the employee is willingly paid less after receiving the bartered item.

If the bread was simply given to the employee with no reduction in income, then it is as if the utility received no connection fee from the ratepayer and thus cannot account for any reduction of expenses.

EXAMPLE: UTILITY CALCULATING CASH AVAILABLE FOR DEBT SERVICE

Cash Available for Debt Service

The final step is to subtract Total Cash Expenses from Regular Income to arrive at CADS:

Regular Income
− Total Cash Expenses
= Cash Available for Debt Service

11

Example: Utility Calculating CADS (cont.)

The Clean Water Utility Company's (CWUC) income and expenses for 2004 are listed on the following slide and they are considering financing the construction of a new disinfection building.

- Part A: What is the amount of regular income?
- Part B: What is the total amount of cash expenses?
- Part C: How much cash is available for debt service?

16

Example: Utility Calculating CADS (cont.)

Income	Description	Amount (USD)
Tariffs	Monthly charges paid by all 5,000 regular customers for water service Monthly charge is (USD) 15 per user per month	900,000
New Connection fees	300 new connections in 2004 down from 900 in 2003	150,000
Interest on a cash in bank	Interest on cash from the land sale	25,000
Sale of land	Parcel of land that had been used for storage of chemicals but will not be needed once new disinfection building is complete.	300,000
Fee for service	CWUC responded to an emergency request from a neighboring utility (NWUC) to provide staff and equipment to repair a broken water main. In addition to compensating CWUC for it expenses, NWUC paid a %10 fee.	15,000
Electricity	Ongoing agreement to get 2 months of free electricity in exchange for water. Average monthly electric bill is 2,500	5,000
Meat	100 lbs. of beef was bartered for a years worth of water by one customer	1,000
Total Income		1,396,000

Example: Utility Calculating CADS (cont.)

Expenses	Description	Amount (USD)
Salaries	Permanent fulltime staff of 50	1,000,000
	Temporary staff of 15 to install 300 new connections	150,000
Chemicals and supplies		25,000
Electricity	10 months of electricity	25,000
Taxes and insurance		7,500
Depreciation	Average annual depreciation for all buildings and equipment. 2004 was the fifth year in an assumed 30 year life	12,000
Lunch for CWUC employees	CWUC provides a free lunch everyday to all of its employees	10,000
Spare parts	Normal year for replacement and repair	6,500
Total Expenses		1,236,000

18

The Clean Water Utility Company (CWUC) has the following income and expenses for 2019 and they are contemplating financing the construction of a new disinfection system. How much cash is available for debt service?

Income	Description	Amount (USD)
Tariffs	Monthly charges paid by all 6,000 regular customers for water service Monthly charge is (USD) 15 per user per month	900,000
New connection fees	300 new connections in 2019 down from 900 in 2018	150,000
Interest	Interest on cash in bank	25,000
Sale of land	Parcel of land that had been used for storage of chemicals but will not be needed once new disinfection system is complete.	300,000
Fee for service	CWUC responded to an emergency request from a neighboring utility (NWUC) to provide staff and equipment to repair a broken water main. In addition to compensating CWUC for it expenses, NWUC paid a 10% fee.	15,000
Electricity	Ongoing agreement to get two months of free electricity in exchange for water. Average monthly electric bill is 2,500	5,000

(Continued)

(Cont.)

Income	Description	Amount (USD)
Meat	100 lbs. of beef was bartered for a year's worth of water by one customer	1,000
Total Income		1,396,000

Expenses		
Salaries	Permanent fulltime staff of ten	200,000
	Temporary staff of 15 to install 300 new connections	150,000
Chemicals and supplies		25,000
Electricity	Ten months of electricity	25,000
Taxes and insurance		7,500
Depreciation	Average annual depreciation for all buildings and equipment. 2019 was the fifth year in an assumed 30 year life	12,000
Lunch for CWUC employees	CWUC provides a free lunch everyday to all of its employees	10,000
Spare parts	Normal year for replacement and repair	6,500
Total Expenses		436,000

Since Cash Available for Debt Service is equal to Regular Income minus Cash Expenses we need to calculate Regular Income and Cash Expenses.

Regular Income = Total Income – Non-Recurring Income
Total Income is 1,396,000

In this example Non-Recurring Income includes:

New connection fees (It appears that the number of new connections is decreasing and may be very small in the future)	150,000
Sale of land (This is one time event and will not provide income in the future)	300,000
Fee for service (Although the Utility may be called on in the future to help neighboring utilities this fee is not a predictable source of income)	15,000
Total Non-Recurring Income	465,000

CALCULATING REGULAR INCOME

Total Income	1,396,000
Non-Recurring Income	465,000
Regular Income	931,000

CALCULATING CASH EXPENSES

The only Non-Cash Expense in the above example is the 12,000 for Depreciation. Therefore, Cash Expenses equal 424,000.

Ex: Utility Calculating CADS (cont.)

Since Cash Available for Debt Service is equal to Regular Income minus Total Cash Expenses, we need to calculate Regular Income and Total Cash Expenses.

Part A:

Regular income = Total Income – Non-recurring income

19

Ex: Utility Calculating CADS (cont.)

Part A (cont.):
Total Income is 1,396,000
In this example, Non-recurring income includes:

- New connection fees → 150,000
(It appears that the number of new connections is decreasing and may be very small in the future)
- Sale of land → 300,000
(This is one time event and will not provide income in the future)
- Fee for Service → 15,000
(Although the Utility may be called on in the future to help neighboring utilities this fee is not a predictable source of income)

Total Non-recurring Income = 465,000
Regular Income = Total Income – Non-Recurring Income
Regular Income = 1,396,000 – 465,000 = 931,000

20

Ex: Utility Calculating CADS (cont.)

Part B:

Total Cash Expenses = Total Expenses – Non-Cash Expenses

CWUC must account for the goods it has bartered. Since CWUC bartered 2 months of water for 2 months of electricity, the annual electricity expense for CWUC is now for 10 months instead of a full year.

In addition, a year's worth of water was bartered for 100 lbs. of beef valued at (USD) 1,000.

We can assume this beef was given to willing employees in lieu of (USD) 1,000 of their cash income.

(USD) 1,000 will need to be subtracted from CWUC's salary expense:

Total Salary Expense = 1,150,000 – 1,000 = 1,149,000

21

Ex: Utility Calculating CADS (cont.)

Part B (cont.):

Total Expenses = 1,236,000

Non-Cash Expenses include: Depreciation = 12,000

Total Cash Expenses = 1,236,000 – 12,000 – 1000 = 1,223,000

(USD) 1000 will be subtracted from this year's salary expense

22

Ex: Utility Calculating CADS (cont.)

Part C:

CADS = Regular Income – Total Cash Expenses

Regular Income = 931,000

Total Cash Expenses = 1,223,000

CADS = 931,000 – 1,223,000 = (292,000)

23

CASH AVAILABLE FOR DEBT SERVICE

As noted above, Cash Available for Debt Service equals Regular Income less Cash Expenses.

Regular Income	931,000
Cash Expenses	424,000
Cash Available for Debt Service	507,000

Questions and Answers

Question #1:

If the utility barters off a piece of old machinery for several barrels of chemicals and then sells the chemicals for (USD) 700. The utility then transports the chemicals by way of a company vehicle to a buyer located eight hours away. The cost to the utility is (USD) 100 in cash payments for the driver of the vehicle and (USD) 80 for the fuel.

What is the amount of revenue from the sale of the old machinery?

How much profit is from the chemicals?

What is the amount of cash payments from the sale of the old machinery?

Question #2:

The Clean Water Utility Company (CWUC) has the following income and expenses for 2005 and they are contemplating financing the construction of a new disinfection building.

Part A: What is the amount of Regular Income?

Part B: What is the amount of total Cash Expenses?

Part C: How much cash is available for Debt Service?

Income	Description	Amount (USD)
Tariffs	Monthly charges paid by all 4,000 regular customers for water service Monthly charge is (USD) 11 per user per month	528,000
New connection fees	200 new connections in 2005 down from 700 in 2004	75,000
Interest on cash in bank	Interest on cash from the sale of machinery	20,000
Sale of land	Parcel of land that had been used for storage of chemicals but will not be needed once new disinfection building is complete	350,000

(*Continued*)

(Cont.)

Income	Description	Amount (USD)
Fee for service	CWUC responded to an emergency request from a neighboring utility (NWUC) to provide staff and equipment to repair a broken water main. In addition to compensating CWUC for its expenses, NWUC paid a 15% fee.	25,000
Electricity	Ongoing agreement to get four months of free electricity in exchange for water. Average monthly electric bill is 3,000	12,000
Meat	150 lbs. of beef was bartered for a year's worth of water by one customer	5,000
TOTAL INCOME		1,015,000
Expenses		
Salaries	Permanent fulltime staff of 40	875,000
Chemicals and supplies	300 barrels at (USD) 75 per barrel	22,500
Electricity	Eight months of electricity	20,000
Depreciation	Average annual depreciation for all buildings and equipment in 2005 was the fifth year in an assumed 30 year life	12,000
Lunch for CWUC employees	CWUC provides a free lunch everyday to all of its employees	11,000
Spare parts	Normal year for replacement and repair	7,500
TOTAL EXPENSES		957,000

Question #3:

$$(\text{Total Cash Income} - \text{Non-Recurring Cash Income}) - (\text{Total Expenses} - \text{Non-Cash Expenses}) = ???$$

Question #4:

$$\text{Regular Cash Income} + \text{Non-Recurring Cash Income} = ???$$

Answer #1:

(USD) 700 is the revenue from the sale.

The profit from the sale of the chemicals is (USD) 700–180 or (USD) 520.

Thus (USD) 520 is considered cash payment from the sale of the old machinery.

Answer #2:

Part A:

Since Cash Available for Debt Service is equal to Regular Income minus Total Cash Expenses, we need to calculate Regular Income and Total Cash Expenses.

Regular Income = Total Income – Non-Recurring Income

Total Income is 1,015,000

In this example, Non-Recurring Income includes:

New connection fees (It appears that the number of new connections is decreasing and may be very small in the future)	75,000
Sale of land (This is one time event and will not provide income in the future)	350,000
Fee for service (Although the Utility may be called on in the future to help neighboring utilities this fee is not a predictable source of income)	25,000
Total Non-Recurring Income	450

Total Income = 1,015,000
Non-Recurring Income = 450,000
Regular Income = 1,015,000–450,000 = 565,000

Part B:

Total Cash Expenses = Total Expenses – Non-Cash Expenses

Note: CWUC must account for the goods it has bartered. Since CWUC bartered four months of water for four months of electricity, the annual electricity expense for CWUC is now for eight months instead of a full year.

In addition, a year's worth of water was bartered for 150 lbs. of beef valued at (USD) 5,000.

We can assume this beef was given to willing employees in lieu of (USD) 5,000 of their cash income.

(USD) 5,000 will need to be subtracted from CWUC's salary expense:

$$\text{Total Salary Expense} = 875,000 - 5,000 = 870,000$$

Total Expenses is 948,000

Non-Cash Expenses include: Depreciation = 12,000

$$\text{Total Cash Expenses} = 948,000 - 12,000 - 5,000 \;= 931,000$$

Note: The above (USD) 5,000 was subtracted from this year's salary expense.

Part C:

Cash Available for Debt Service (CADS) = Regular Income – Total Cash Expenses
CADS = 565,000 – 931,000 = (366,000)

Answer #3:

Cash Available for Debt Service

Answer #4:

Total Cash Income

Example: Utility Calculating Cash Available for Debt Service

Income	**Description**	**Amount (USD)**
Tariffs	Monthly charges paid by all 5,000 regular customers for water service Monthly charge is (USD) 15 per user per month	900,000
New connection fees	300 new connections in 2004 down from 900 in 2003	150,000
Interest on a cash in bank	Interest on cash from the land sale	25,000
Sale of land	Parcel of land that had been used for storage of chemicals but will not be needed once new disinfection building is complete.	300,000
Fee for service	CWUC responded to an emergency request from a neighboring utility (NWUC) to provide staff and equipment to repair a broken water main. In addition to compensating CWUC for it expenses, NWUC paid a 10% fee.	15,000
Electricity	Ongoing agreement to get two months of free electricity in exchange for water. Average monthly electric bill is 2,500	5,000
meat	100 lbs. of beef was bartered for a year's worth of water by one customer	1,000
Total Income		1,396,000

(*Continued*)

(Cont.)

Expenses	Description	Amount (USD)
Salaries	Permanent fulltime staff of 50.	1,000,000
	Temporary staff of 15 to install 300 new connections.	150,000
Chemicals and supplies		25,000
Electricity	Ten months of electricity.	25,000
Taxes and insurance		7,500
Depreciation	Average annual depreciation for all buildings and equipment. 2004 was the fifth year in an assumed 30 year life.	12,000
Lunch for CWUC employees	CWUC provides a free lunch everyday to all of its employees.	10,000
Spare parts	Normal year for replacement and repair.	6,500
Total Expenses		1,236,000

The Clean Water Utility Company (CWUC) has the following income and expenses for 2004 and they are contemplating financing the construction of a new disinfection building. How much cash is available for Debt Service?

Since Cash Available for Debt Service is equal to Regular Income minus Cash Expenses we need to calculate Regular Income and Cash Expenses.

$$\text{Regular Income} = \text{Total Income} - \text{Non-Recurring Income}$$
$$\text{Total Income is } 1,396,000$$

In this example Non-Recurring Income includes:

New connection fees (It appears that the number of new connections is decreasing and may be very small in the future)	150,000
Sale of land (This is one time event and will not provide income in the future)	300,000
Fee for service (Although the Utility may be called on in the future to help neighboring utilities this fee is not a predictable source of income)	15,000
Total Non-Recurring Income	465,000

Total Income	1,396,000
Non-Recurring Income	465,000
Regular Income	

3 Maximizing Cash Available for Debt Service

INTRODUCTION

In the preceding chapter, Measuring Income, gross income was discussed along with expenses. The income remaining after payment of all cash expenses from recurring income was defined as *Cash Available for Debt Service (CADS)*.

In this chapter, we depart from purely financial considerations to take a practical look at how to maximize CADS.

Why?

Because, the amount of CADS determines how much debt a utility can take on. And, unless there is grant money available, the amount of debt will be equal to the size of the project, which the utility can undertake. So, it is important to maximize CADS to enable the utility to take on larger projects, if necessary.

Introduction

- Cash Available for Debt Service (CADS): Income remaining after payment of all cash expenses from recurring income
- Amount of CADS determines how much debt a utility can take on
- Maximization of CADS enables the utility to take on larger projects

2

CHAPTER CONTENTS

- Tariffs
- Subsidies
- Billings and Collections
- Theft
- Energy Costs
- Chemical Costs
- Labor Costs
- Modernization of Facilities
- Staff Skills

This chapter presents a series of strategies to maximize the net operating income (which is another word for CADS) of water utilities by maximizing their revenues and minimizing their expenditures.

Maximizing Net Water Utility Revenues

Goal: To have sufficient funds available to operate the system on a sustainable basis, **not** to make a profit.

There are two approaches to address the problem of maximizing net water utility revenues:

- Maximization of Income
- Minimization of Expenses

3

The purpose of this effort is to maximize net water utility revenues so that funds will be sufficient to pay for ongoing operation and maintenance costs and to provide at least some cash to support debt service on projects to improve or expand systems.

Maximizing CADS is not a problem unique to any particular country. Indeed, in the most developed countries, water utility operators still struggle to keep tariff income above the ever-rising cost of operation.

There are over 55,000 water systems in the United States. Small community systems belong to their respective state rural water association, which are affiliated with the National Rural Water Association (NRWA). Larger systems belong to the American Water Works Association (AWWA). Both of these associations have developed programs, materials and procedures to help their members with tariffs, targeted subsidy and other revenue problems, as well as with cost-reduction efforts. These can be made available to municipal utility operators in foreign companies through the Office of International Affairs of the US Environmental Protection Administration (USEPA).

The goal of any initiative to maximize net utility revenues is not to make a profit, but rather to have sufficient funds available to operate the system on a sustainable basis so that the user community will have continuing access to adequate supplies of clean, safe water.

There are two approaches to the problem of maximizing net water utility revenues. First, is to maximize income. Second, is to minimize expenses.

MAXIMIZATION OF INCOME

There are five elements of the initiative to maximize income.

Maximization of Income

Five elements:

- Increase tariffs
- Establish consumption-based tariffs
- Replace general subsidies with targeted subsidies
- Modernize billing & collections
- Eliminate theft

4

1. *Increase tariffs*: Increasing tariffs is always politically difficult. It is especially difficult to raise tariffs if the service has been poor and/or intermittent. But strategies can be developed to make it politically possible. The simplest and most direct of these is to organize a small demonstration project so that citizens can see the benefits of improved service. For example, if contaminated water is the problem, a certain small geographic area could be furnished with clean, bottled water for 30 days and the testimony from those people concerning the decrease in waterborne disease could be used to convince the rest of the system users to pay higher tariffs for system-wide safe water.

Maximization of Income

I. Increase tariffs

- Politically difficult, especially if past service has been poor
- Simplest, most direct strategy:

Organize small projects demonstrating benefits of improved service

5

2. *Establish consumption-based tariffs*: Fair and equitable tariff schedules can be devised to allocate cost based on use. As discussed previously, no one likes to pay for water. This is especially so if people believe they are being unfairly charged. For example, let us take a household of five people that uses 1,000 liters/day and a food processing plant nearby that uses 100,000 liters/day. If both are charged the same amount of money, the householders will be justifiably angry.

   Designing tariffs that increase in accord with consumption is quite easy if each outlet is metered. But even where there are no meters, parameters can be introduced (such as average per capita consumption) that can serve as the basis for incremental tariffs. (Tariff structure is more fully discussed in Chapter 10 – Tariff Design.) In addition, EPA's Environmental Finance Center at Boise State University in Idaho has developed a computerized tariff setting program that can be used even in very small communities and is available from EPA.

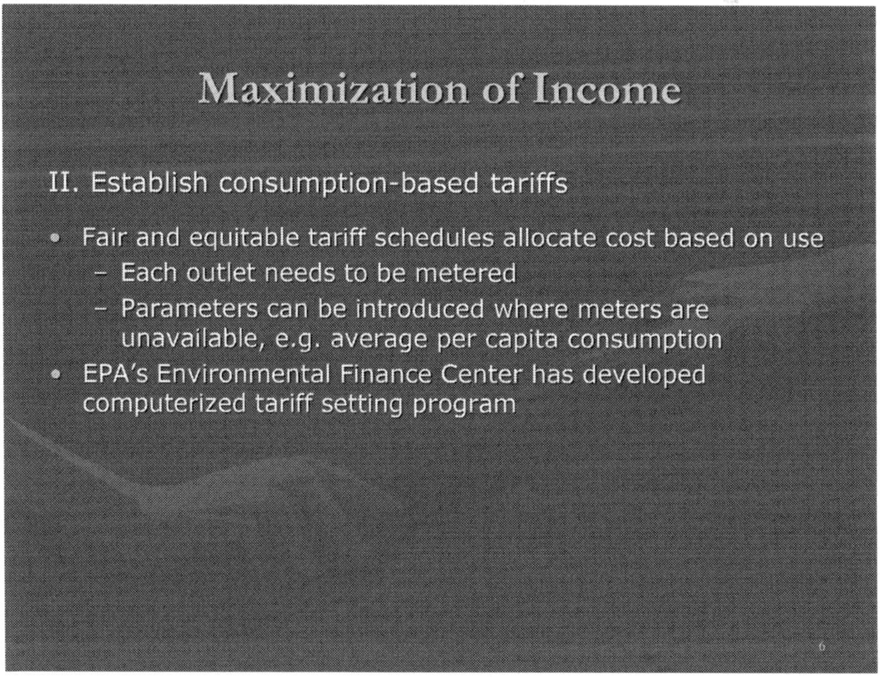

Maximization of Income

II. Establish consumption-based tariffs

- Fair and equitable tariff schedules allocate cost based on use
  - Each outlet needs to be metered
  - Parameters can be introduced where meters are unavailable, e.g. average per capita consumption
- EPA's Environmental Finance Center has developed computerized tariff setting program

6

3. *Replace general subsidies with targeted subsidies*: Widows, pensioners, the disabled, and others living on low, fixed incomes, as well as the poor, must be provided for. But no one else. If the water tariff exceeds certain users' abilities to pay, then the utility (or the community at large) must make provision for paying their share of the cost of operating the system. Low tariffs must be replaced with full cost recovery tariffs; so, appropriate exemptions or other forms of targeted subsidies must be established for the truly needy.

Maximization of Income

III. Replace general subsidies with targeted subsidies

- Replace low tariffs with full cost recovery tariffs
- Establish exemptions or other forms of targeted subsidies for truly needy

7

4. *Modernize billing and collections*: Modern methods of tracking users must be introduced. Their amount of use (if applicable) must be determined. Each user must be billed the correct amount in a timely manner. Accurate and publicly available collection records must be kept. Finally, effective collection mechanisms must be created for normal as well as delinquent accounts.

Maximization of Income

IV. Modernize billing & collections

- Introduce modern methods of tracking users
- If applicable, determine amount of use
- Bill user correct amount in a timely manner
- Keep collection records accurate and publicly available
- Create effective collection mechanisms for normal and delinquent accounts

8

5. *Eliminate theft*: Illegal, non-paying users of the system must be identified, offered the opportunity to become regular billed customers, or disconnected from the system. This process begins with a modern method of tracking each individual user.

Maximization of Income

V. Eliminate theft

- Develop modern method of tracking each user
- Identify illegal, non-paying users
- Offer illegal users opportunity to become billed customers, otherwise disconnect from system

9

The above five initiatives can maximize the income a utility receives from its operations. Maximizing utility income is the first important step in maximizing a utility's CADS.

MINIMIZATION OF EXPENSES

There are five major areas where significant cost reductions are most likely to be available.

Minimization of Expenses

Five elements:

- Reduction of energy costs
- Reduction of chemical costs
- Maximum allocation of labor
- Modernization of plant & equipment
- Amelioration of staff skills

10

1. *Reduction of energy costs*: Many water systems have enormous energy costs. These are most often due to antiquated, inefficient machinery and equipment, or to massive leakages that cause huge amounts of energy to be used to keep water flowing to end users. Energy use must be audited and reduced in accord with audit findings. For example, if energy is less expensive during the night, and water storage facilities are available, the electrical pumps should be turned on at night to fill the storage facilities and use very little during the more expensive daylight periods.

Minimization of Expenses

I. Reduction of energy costs

- Inefficient machinery and massive leakages result in water systems having high energy costs
- Energy must be audited and reduced in accord with findings

11

2. *Reduction of chemical costs*: Many water systems have unusually large costs for water treatment chemicals. There are strategies that can be adopted to reduce these costs. One cause of high chemical costs are multiple treatments because of recontamination. Measures must be implemented to assure that once water is decontaminated, it remains so.

Minimization of Expenses

II. Reduction of chemical costs

- Water treatment chemicals are costly
- Recontamination results in multiple treatments
- Implement measures to ensure decontaminated materials remain so

12

3. *Maximum allocation of labor*: Many systems (in all countries) are overstaffed. Staff reduction can sometimes pose a political problem. But, often there are opportunities to reallocate labor into more productive areas such as billing and collections.

Minimization of Expenses

III. Maximum allocation of labor

- Systems overstaffed
- Staff reduction can result in political problems
- Search for opportunities to reallocate labor into more productive areas

13

4. *Modernization of plant and equipment*: Many systems could reduce operating and especially maintenance costs through the replacement of antiquated, inefficient machinery and equipment. When including replacement machinery and equipment in a new project, care must be taken to replace the old inefficient operating cost with an accurate estimate of the new, lower, more efficient costs. This will have the concomitant effect of increasing CADS.

Minimization of Expenses

IV. Modernization of plant & equipment

- Replace inefficient machinery and equipment
- Carefully replace old operating costs with accurate estimates of new, lower costs
- CADS will increase as a result

14

5. *Amelioration of staff skills*: Exposing staff to modern utility management methods through the furnishing of training materials and presentations at workshops and seminars can greatly improve the efficiency of systems and help reduce their costs.

Minimization of Expenses

V. Amelioration of staff skills

Improve efficiency of systems and reduce costs by exposing staff to:

- Modern utility management methods
- Training materials
- Presentations at workshops and seminars

15

For urban communities with established water utilities, initiatives to enhance net system revenues through innovative income generating strategies and aggressive cost cutting measures can truly result in safer, cleaner water and a more healthful life for the residents of those communities.

The first five initiatives above can maximize the income a utility receives from its operations. Below those are another set of five initiatives that can minimize a utility's cash expenses. Taken together, these initiatives can maximize CADS.

Conclusion

Innovative income generating strategies and aggressive cost cutting measures can result in safer, cleaner water and a more healthful life for the residents of those communities with established water utilities.

16

Questions and Answers

Question #1:
What is the goal of any initiative focusing on maximizing net utility revenues?

Question #2:

When the maximization of the income a utility receives from its operations is combined with the minimization of a utility's cash expenses, what is the desired outcome referred to?

Question #3:

If a household of four people use 800 liters/day and a nearby food processing plant uses 90,000 liters/day, what would be the most appropriate method to allocate the cost of water consumed by these two groups?

Question #4:

What are two factors contributing to the enormous energy costs of operating a water system?

Question #5:

What is one major cause of high chemical costs associated with water systems?

Answer #1:

To have sufficient funds available to *operate the system on a sustainable basis, not to make a profit*, so that the user community will have continuing access to adequate supplies of clean, safe water.

Answer #2:

Cash Available for Debt Service (CADS).

Answer #3:

Establish consumption-based tariffs.

Answer #4:

Inefficient machinery and massive leakages.

Answer #5:

Recontamination.

4 Loan Basics

INTRODUCTION

A loan is defined as the use of money belonging to another person, or institution, for a definite period of time at which it is returned. Borrowing money is economically sound and prudent for many reasons.

Introduction

- *Loan*: Use of money belonging to another person, or institution, for a definite period of time at which it is returned.
- Borrower is legally obligated to repay the loan
- Borrower must be concerned with the amount of repayment

2

Once money is borrowed, the borrower is legally obligated to repay the loan. Therefore, borrowers must be concerned with the amount of the repayment. The amount of the repayment will depend on four factors.

First, the amount of repayment will directly depend on the amount borrowed, which is called the “Principal”. Second, the amount of the repayment will depend on the period of time for which it is borrowed. This time period is called the “Term”. Third, the amount of repayment will depend on the amount of money the lender charges the borrower for the use of the money. These charges are called “interest”, or “interest payments”. Interest payments are most often expressed as a fixed percentage of the unpaid principal. This percentage is called the "interest rate". Fourth, repayments may also be affected by certain initial or periodic charges which can be made by the lender (and agreed to by the borrower).

Introduction

Amount of repayment depends on four factors:

- *Principal*: Amount borrowed
- *Term*: Period of time amount is borrowed
- *Interest*: Amount of money the user charges for the use of the money. Interest payment is often expressed as a fixed percentage of the unpaid principal, called the *interest rate*.
- *Initial or Periodic Charges*: Borrower must agree on charges prior to receiving loan

3

Finally, perhaps the most difficult, but most important, concept in finance is that, because of inflation, the value of money changes over time. This change in value over time must be measured accurately to determine the true cost of loan repayments. This concept is called the "Time-Value" of Money. Loan options of different rates and terms can be compared only by using the Time-Value of Money. The Time-Value of Money will be discussed in the next module.

Introduction

Final factor in determining the true cost of loan payments:

Time Value of Money: Change in value over time due to inflation. Allows for loan options of different rates and terms to be compared.

4

CHAPTER CONTENTS

- Reasons to Borrow
  - Availability of Money
  - "Life" of the Asset
  - Fairness
- Elements of a Loan
  - Principal
  - Term
  - Rate
  - Relationship between Rate and Term
  - Fees or Charges
- Time Value of Money
  - Discounting
  - Compounding
  - Choosing a Discount Rate
    - Reinvestment Rate
    - Inflation Rate

REASONS TO BORROW

The first and most basic reason to borrow is unavailability of money: the borrower does not have enough money to pay for the project in cash. Nor is the borrower able to obtain money for free. No gifts are available from rich uncles; no grants are available from government agencies.

Reasons to Borrow

I. Unavailability of Money

- Borrower has insufficient money to pay for the project in cash
- Borrower is unable to obtain money for free

5

The second reason to borrow concerns the purpose for which the money will be spent. If the purpose of the money is to buy things that will be consumed immediately or in the near future – such as food or supplies – then it should be paid for in cash. Things that will be consumed immediately or in the near future are called "short-term assets". But if the things to be purchased with the money will be used over many years – such as a vehicle or a house – it is both prudent and economically sound to pay for them over the same approximate period of time for which they will be used. The period for which assets can be used is called their "service life". Assets with service lives of more than one year are called "long-term assets".

The principle of paying for an asset over its service life can be illustrated by the case of a driver who has no money but wishes to purchase a truck to earn money by delivering goods. If the truck has a service life of five years, costs $500, and will make the driver a profit of $250 per year. Then, it would be appropriate for the driver to borrow the $500 to purchase the truck. If there were no interest payments, he could pay $100 per year for five years, and still make $150 per year profit. He could also pay 10% interest. In this case, his annual repayments would be $150, $140, $130, $120 and $110; which would mean that his profits would still be $100, $110, $120, $130 and $140, respectively.

Reasons to Borrow

II. "Service Life" of the Asset

- Period for which assets can be used
- Two types:
  - Short-term: Assets consumed immediately or in the near future, paid for in cash, e.g. food
  - Long-term: Service life over one year, paid for over the period of time for which they are used, e.g. vehicle

6

The third reason for borrowing is fairness. This concept is especially relevant to public projects, or projects that will benefit a large number of people over many years. Let us use the example of a new town water system with a service life of 30 years. In this case, many different people will drink the water over the 30-year service life. People will emigrate away from the town; and people will immigrate to the town. People using it today will have children and even grandchildren who will drink. It is both fair and appropriate that all of the people who drink pay for part of the system at the time that they use it.

Reasons to Borrow

III. Fairness

- Especially relevant to public projects, projects benefiting a large number of people over many years
- Example:

  New town water system has service life of 30 years. Many townspeople will drink from it over time. Therefore, it is both fair and appropriate for all people who drink from it to pay for part of the system at the time of use.

7

ELEMENTS OF A LOAN

Loans have three variables:

- *Principal* is the actual dollar amount of the loan. *Interest* is the cost of taking a loan. It is a function of the interest *rate* and the principal balance. The interest due and payable in any year is always calculated as follows:

  Interest for the Current Year = Outstanding Principal Balance of the Loan ∗ Interest Rate

- *Term* is the number of years during which loan principal is still "outstanding" (i.e., at least a portion of it has not yet been paid).

There is a relationship between rate and term. Rates are often affected by time. This will be more fully discussed below.

Fees and charges also affect loans; but they do so by either affecting rate or principal.

Elements of a Loan

Loans have three variables:

- *Principal*: Actual dollar amount of loan
- *Term*: Number of years loan principal is still "outstanding"
- *Interest*:
  - Function of the interest *rate* and the principal balance.
  - Interest due and payable in any year is calculated as follows:

Interest for the current year =
Outstanding principal balance of the loan * Interest rate

- *Fees and charges*: Affect either rate or principal

8

PRINCIPAL

Using a town water utility again as an example, let us say that a new pumping station costs $40,000 and a few kilometers of pipe another $60,000. Let us say that the engineering costs $6,000. In this case, the total project cost is $106,000. Let us say that the utility has $11,000 of cash on hand. In such case, the utility might decide to borrow $100,000, pay the $6,000 engineering fees in cash and keep $5,000 in cash in case of emergencies. The $100,000 borrowed is the principal of the loan. It is not the same as the project cost. Here, the project cost is $106,000; and the loan principal is $100,000. In borrowing the money, the water system would agree to pay back the $100,000 principal plus an additional amount of money, the interest, each year, over a period of years called the "term".

Elements of a Loan: Principal

Example:

- New pumping station: $40,000
- Few Kilometers of pipe: $60,000
- Engineering: $6,000
- Total project cost = $106,000

9

Elements of a Loan: Principal

Example (cont.):

- Utility has $11,000 cash on hand
- Utility decides to borrow $100,000
- Utility pays the $6,000 engineering fees in cash
- Utility keeps remaining $5,000 in case of an emergency
- Principal of loan = $100,000
- Utility agrees to pay back the $100,000 principal plus interest, each year, over a period of years called the "term"

10

TERM

- *Maturity* – the time at which an installment of principal is repaid. Thus the "final" maturity is synonymous with term.
- *Amortization schedule* – amortization is the process of repaying principal, thus the amortization schedule is a list of maturity dates and amounts of principal due and payable on those dates. This is essentially the same as the *Annual Debt Service Payment Schedules* discussed in Chapter 5.

Elements of a Loan:
Term

- *Maturity*: Time an installment of principal is repaid, final maturity is synonymous with term
- *Amortization*: Process of repaying principal
- *Amortization Schedule*: List of maturity dates and amounts of principal due and payable on those dates

11

Term is very important. The length of the term has a great impact on the amount of each loan repayment each year. For example, assume that a water system borrows $100,000 at a 0% interest rate and agrees to pay the loan back in equal installments.

Elements of a Loan: Term

- The length of the term has a great impact on the amount of loan repayment each year.
- For example: Assume a water system borrows $100,000 at a 0% interest rate and agrees to pay back the interest loan in equal installments.

12

Here is what the annual payments would be depending on the term:

Term	Annual Payment
1	$100,000
2	$50,000
5	$20,000
10	$10,000
20	$5,000
25	$4,000
40	$2,500

Elements of a Loan: Term

Example (cont.)

The table below illustrates what the annual payments would be depending on the term:

Table I

Term	Annual Payment ($)
1	100,000
2	50,000
5	20,000
10	10,000
20	5,000
25	4,000
40	2,500

Assuming the rate, principal and other factors are the same, *the longer the term, the lower the annual debt service payment.* If a system can only afford to pay $10,000/year, a 10 year term will be needed on their loan.

13

Assuming the rate, principal, and other factors are the same then, *the longer the term the lower the annual debt service payment*. This means that if a system can only afford to pay $10,000 per year, they will need a ten year term on their loan.

RATE

Rate is one of many factors which determine the annual and total cost of financing a project. In considering rates, there are three issues to discuss:

- How rates are set.
- Interest and annual debt service payment schedules.
- Understanding the relationship between rate and time.

Elements of a Loan:
Rate

- Rate is one of many factors determining the annual and total cost of financing a project
- Three issues to discuss:
  - How rates are set
  - Interest and annual debt service payment schedules
  - Understanding the relationship between rate and time

14

How Rates Are Set

There are four basic components that make up rates:

- *Inflation* – as mentioned earlier in this chapter, the value of money changes over time. A lender wants the principal paid back to have the same buying power as it did when it was lent, so if the rate of inflation is estimated at 5% (next year it will costs $105 to buy something that costs $100 today), then the minimum interest rate is 5%.

How Rates are Set

Basic Rate Components:

1) *Inflation:*

- The value of money changes over time
- A lender wants the principal paid back to have the same buying power as it did when it was lent
- For example, if the rate of inflation is estimated at 5% (next year, it will cost $105 to buy something that costs $100 today), as a result the minimum interest rate is 5%

15

- *Risk* – the more likely the borrower is to default (not repay the loan) the higher the risk to the lender. So if the lender calculates the average loss from default is 3% of the principal, then the lender will increase the rate by 3% to recover these expected losses. In this ongoing example, the rate of inflation is 5%, plus 3% of risk recovery, which equals an 8% rate without the lenders actually making any profit for themselves.

How Rates are Set

Basic Rate Components (cont.):

2) *Risk:*

- The higher the borrower's likelihood to default, the higher the risk to the lender
- For example, if the lender calculates the average loss from default to be 3% of the principal, the lender will increase the rate by 3% to recover these expected losses
- If inflation is 5%, plus 3% risk recovery, 8% will now be the rate without the lenders making any profit

16

- *Cost* – lenders are usually businesses that have costs, such as staff and buildings. These costs must also be factored into the interest rate that the lender will charge the borrower. If the lender calculates that his costs (over all the loans he makes to all borrowers) is 1%, he will add another 1% to the rate, for a total of 9%.

How Rates are Set

Basic Rate Components (cont.):

3) *Cost:*

- Lenders are usually businesses that have costs, e.g. staff
- Costs must be factored into the interest rate the lender will charge the borrower
- For example, if the lender's calculated costs (over all loans made to borrowers) is 1%, he will add 1% to the total rate
- If inflation is 5%, recovery costs are 3% and the lender's costs are 1%, the total rate is now 9%

17

- *Profit* – lenders will also want to make a profit. In this case, the lender will want to make a 3% profit, as a minimum. This means that the lowest rate the lender will charge the borrower is 12%. *However*, if a lender can get a rate of return of 14% from investing the money elsewhere (at an equal risk), then this *opportunity cost must also be considered*. In such case, the lender will charge the borrower 14% and make a 5% profit.

How Rates are Set

Basic Rate Components (cont.):

4) *Profit:*

- Assume the lender wants to make a 3% profit, as a minimum
- The lowest rate the lender will need to charge the borrower will be 12% (5% inflation, 3% recovery, 1% lender's cost and 3% profit)
- If the lender can get a rate of return of 14% from investing the money elsewhere (at an equal risk), then this opportunity cost must also be considered. The lender will now charge the borrower 14%, making a 5% profit

18

Interest and Annual Debt Service

Now that the interest rate (14%) is known, we can calculate the *annual debt service* for the *first year* (only), as follows for the different terms:

Term	Annual Principal Payment	Annual Debt Service Payment
1	$100,000	$114,000
2	$50,000	$64,000
5	$20,000	$34,000
10	$10,000	$24,000
20	$5,000	$19,000
25	$4,000	$18,000
40	$2,500	$16,500

Interest and Annual Debt Service

Building on the previous example, assume the interest rate is 14%, the annual debt service for the first year (only) can be calculated as follows for the different terms:

Table II

Term	Annual Principal Payment ($)	Annual Debt Service Payment ($)
1	100,000	114,000
2	50,000	64,000
5	20,000	34,000
10	10,000	24,000
20	5,000	19,000
25	4,000	18,000
40	2,500	16,500

Assuming the term, principal and other factors are the same then, the lower the rate, the lower the annual and total cost. Note: For the terms of 20, 25 and 40 years, the difference between the annual debt services is very low. This means the benefits gained from extending the term are equally low.

19

Assuming the term, principal, and other factors are the same then, the lower the rate the lower the annual and total cost. But note that, for the terms of 20, 25, and 40 years, the difference between the annual debt services is very low. This means that the benefits gained from extending the term are commensurately low.

It is also important to note that *if principal is paid off in level installments, then the annual debt service (or total payment) each year will decline over time.* This is illustrated for a loan with a five year term in the table below.

Year	Principal Balance	Principal Payment	Interest Payment	Total Payment
1	$100,000	$20,000	$14,000	$34,000
2	$80,000	$20,000	$11,200	$31,200
3	$60,000	$20,000	$8,400	$28,400
4	$40,000	$20,000	$5,600	$25,600
5	$20,000	$20,000	$2,800	$22,800

Interest and Annual Debt Service

If the principal is paid off in level installments, the annual debt service (or total payment) each year will decline over time. This is illustrated for a loan with a five year term in Table III below:

Table III

Year	Principal Balance ($)	Principal Payment ($)	Interest Payment ($)	Total Payment ($)
1	100,000	20,000	14,000	34,000
2	80,000	20,000	11,200	31,200
3	60,000	20,000	8,400	28,400
4	40,000	20,000	5,600	25,600
5	20,000	20,000	2,800	22,800

20

THE RELATIONSHIP BETWEEN TIME AND RATE

As noted above, inflation is a major component of rate. However, the rate of inflation is an historical number, i.e., it is known only for past years. Lenders must, therefore, estimate future rates of inflation. Typically, rates of inflation for the near future will be estimated at rates very close to immediate past rates. Beyond two or three years, however, lenders may want to include a margin of safety in case the rate of inflation increases. In such case, they will charge more interest (higher rates) for the longer term maturities.

Time and Rate Relationship

- Inflation is an historical number, known only for past years
- Lenders must estimate future rates of inflation
- Near future inflation rates will be estimated very close to immediate past rates
- Beyond two or three years, lenders may include a margin of safety in case the rate of inflation increases, charging more interest for the longer term maturities

21

If lenders fear that inflation will steadily increase over time, they might charge rates such as the following:

Year	Rate
1	5%
2	5%
3	5.5%
4	6%
5	7%

Time and Rate Relationship

If lenders fear inflation will steadily increase over time, they may charge rates such as the following:

Year	Rate
1	5%
2	5%
3	5.50%
4	6%
5	7%

When these rates are graphed, they reveal an upward sloping curve, referred to as a *yield curve*. If the yield curve slopes up (meaning lenders fear rising inflation), it is called a *positive yield curve*. If lenders believe inflation is under control and will decline over time, lenders will charge less and the yield curve will slope downward and is called a *negative yield curve*.

22

When these rates are graphed, they reveal an upward sloping curve. Finance specialists refer to this as the *yield curve*. If the yield curve slopes up (meaning that lenders fear rising inflation), it is called a *positive yield curve*. If lenders believe that inflation is under control and that it will decline over time, then the will charge less over time. In such case, the yield curve will slope downward and be called a *negative yield curve*.

What follows is a graph of a positive yield curve which shows rate on the vertical axis and term on the horizontal axis. Generally, the longer the term, the higher the rate. This relationship reflects uncertainty and risk that is inherent over longer term loans.

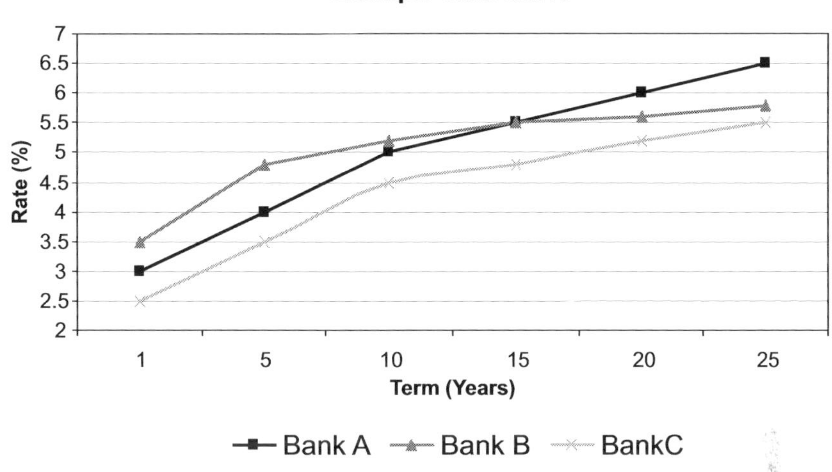

Time and Rate Relationship

Below is a graph of a positive yield curve showing rate on the vertical axis and term on the horizontal axis. Generally, the longer the term, the higher the rate. This relationship reflects uncertainty and risk inherent over longer term loans. Different lenders (represented by the banks) each have their own estimates of inflation and thus will lend at various rates along their own yield curves.

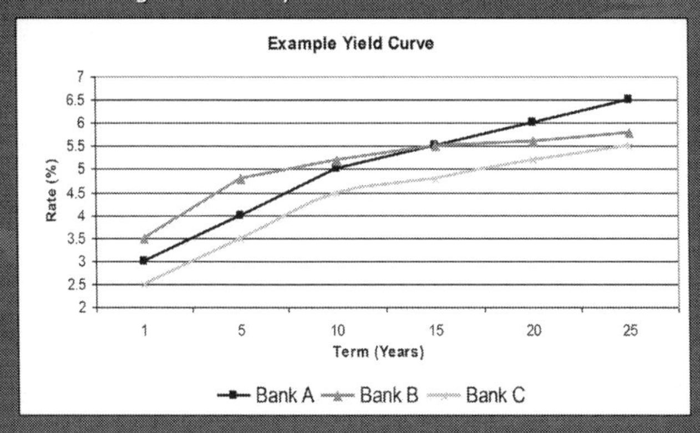

23

As you can see from the above example, different lenders (represented by the banks in the above illustration) each have their own estimates of inflation and thus will lend at various rates along their own yield curves.

FEES AND CHARGES

Lenders often require borrowers to reimburse them directly for certain expenses.

For example, a lender might require a borrower to reimburse its legal fees of $2,000, which are incurred before the loan is made. Using the example above, this $2,000 fee would be added to the original loan principal, for a total loan of $102,000. Thus, the borrower would actually have to repay $102,000, although he only had the use of $100,000 for his project.

Another type of fee might be an "inspection fee". Here, the lender might wish to hire an engineer to inspect the borrower's water system to assure that it is in a good operating condition. Assuming the engineer will charge $1,000 for such inspection, the lender will add $1,000 to the borrower's annual debt service payment.

Another type of fee might be a "servicing fee", which lenders often charge to cover their costs of administering and monitoring the loan. Servicing fees are most often expressed as a percentage of the outstanding balance of the loan. A typical servicing fee might be 0.125% or 0.25%. The lender will calculate this fee each year, based on the outstanding principal balance of the loan, and pass it along to the borrower.

Fees and Charges

- Lenders often require borrowers to reimburse them directly for certain expenses
- Lender may require a borrower to reimburse its legal fees, this fee would be added to the original loan principal
- Inspection fee
  - Lender may want to hirer an engineer to inspect the borrower's water system to assure it is in good operating condition
  - Lender will add cost of engineer to the borrower's annual debt service payment
- Servicing fee
  - Lenders may charge to cover their costs of administering and monitoring the loan
  - Expressed as a percentage of the outstanding balance of the loan
  - Typical servicing fee is between 0.125% and 0.25%

24

Questions and Answers

Question #1:
A driver has no money and is contemplating whether to purchase a truck for the purpose of earning money by delivery goods. The truck has a service life of eight years, costs $1,000, and will make the driver a profit of $500/year. Would it be appropriate for the driver to borrow $1,000 for the truck at a rate of 10% interest over five years? In a table, list the principal balance, principal payment, interest payment, total payment, and profit the driver will make for each year. Assume the principal will be paid off in level installments.

Question #2:
What percentage of profit will the lender make if the rate of inflation is 3%, recovery costs are 4%, costs to the lender are 0.5% and the lender intends on charging the borrower an overall rate of 13%?

Question #3:
A borrower recently accepted a loan for $500,000 from a lender at an interest rate of 10%. If the borrower can afford to pay back the loan at $100,000 - per year, how long of a term will the borrower need to pay off the loan? Use a table to list the principal balance, principal payment, interest payment and total payment for each year.

Question #4:
In general, if the term is longer on a loan, will the interest rate be higher or lower?

Question #5:
If the rate of inflation is estimated at 7% for the following year, what will be the value of $2,000 next year?

Answer #1:
As the table illustrates below, it would be appropriate for the driver to borrow $1,000 for the truck considering the profits each year are in excess of the costs associated with the loan.

Year	Principal Balance	Principal Payment	Interest Payment	Total Payment	Profit
1	$1,000	$200	$100	$300	$200
2	$800	$200	$80	$280	$220
3	$600	$200	$60	$260	$240
4	$400	$200	$40	$240	$260
5	$200	$200	$20	$220	$280

Answer #2:

$$13\% - 3\% - 4\% - 0.5\% = 5.5\%$$

Answer #3:
Eight years.

Year	Principal Balance	Principal Payment	Interest Payment	Total Payment
1	$500,000	$50,000	$50,000	$100,000
2	$450,000	$55,000	$45,000	$100,000
3	$395,000	$60,500	$39,500	$100,000
4	$334,500	$66,500	$33,450	$100,000
5	$268,000	$73,200	$26,800	$100,000
6	$194,800	$80,520	$19,480	$100,000
7	$114,280	$88,572	$11,428	$100,000
8	$25,708	$23,137.20	$2,570.80	$25,708

Answer #4:
Higher. Uncertainty and risk are inherent over longer term loans.

Answer #5:
$2,140.

5 Project Valuation

"The value of money changes over time!"

INTRODUCTION

In Chapters 2 and 4, the concept of income, as it relates to financing projects, and the concepts of rate and term have all been discussed. Now, project valuation will be examined.

Note that this does not refer to the value which an engineer might place on a project in terms of labor and materials; rather here the concept of "financial valuation" will be explored. The idea of "financial valuation" centers on the notion that the value of money actually changes over time; and that to finance a project over time, one must know what is called the "time-value" of money.

Introduction

- *Financial Valuation*: The value of money changes over time and to finance a project over time, one must understand the "time-value" of money
- The value of money changes over time due to inflation
- *Inflation*: overall increase in prices of goods and services in an economy
- As the cost of goods and services increases, the value of a dollar (or any currency) falls

Probably one of the most difficult but important concepts both in finance and economics is that the value of money changes over time. This happens because of inflation. Inflation is the overall increase in prices of goods and services in an economy. As the cost of goods and services increases, the value of a dollar (or any currency) falls commensurately because you are not able to purchase as much with that dollar as you previously could.

This concept is important because it enables us to compare amounts of money at various times.

For example, if you read in a newspaper that a particular country spent $100 million on its army in 1950, and $105 million last year, your instincts will immediately tell you that something is seriously wrong with this situation. You might think that the country must have been at war in 1950, and is at peace today. Or, you will think that the country has, for some other serious reason, drastically reduced its armed forces. In short, even though $105 million is $5 million more than $100 million; you instinctively know that an increase of only 5% over 55 years isn't enough to keep any army in the same fighting condition that it was.

In this example, your intuition is telling you that, although the amount today may be a little bit more, it is not enough. It is telling you that there is a time value to money. It is telling you that the value of money decreases over time.

To fully illustrate this concept, one needs only to consider the cost of a kilo of bread. Unless the government is subsidizing the cost of bread, a kilo of bread costs more today than it did five years ago, usually, considerably more.

Let us say that the cost of a kilo of bread five years ago was $1.00. Let us say that inflation has been 5% each year for the last five years. This means that the price of bread would have increased at a rate of 5% each year. So, one year later (four years ago), the cost of a kilo of bread would have been $1.05.

But, in that year, the cost also increased by 5%. But this time, it was not 5% of $1.00, but rather 5% of $1.05. Economists and finance specialists call this phenomenon "compounding". Thus, three years ago the cost of bread would have been 5% x $1.05, or $1.1025, rounded to $1.10.

Compounding again, two years ago the bread would have cost 5% x $1.1025, or $1.156725, which would be rounded to $1.16.

One year ago, i.e., last year, the same kilo of bread would have cost 5% x $1.156725, or $1.21550625, rounded to $1.22.

Today, the same kilo of bread costs 5% x $1.21550625, or $1.2762815625, rounded to $1.28.

Five years from now – assuming inflation holds at 5% each year – the cost of a kilo of bread will be $1.63.

Introduction

Example 1: Time Value of Money

- The cost of a kilo of bread three years ago was $1.00
- Inflation has been 5% each year for the last three years
- After Year 1 (two years ago), the cost of a kilo of bread would have been $1.00 + 5% = $1.05
- After Year 2 (one year ago), the cost of bread again increased by 5%. The calculation will not be 5% of $1.00, but rather 5% of $1.05, or $1.1025, rounded to $1.10, this phenomenon is referred to as *"compounding"*
- Three years later (today), the cost of bread will be $1.1025 + 5% = $1.156725, rounded to $1.16

Introduction

Example 1 (cont.): Time Value of Money

- Seven years from now, assuming inflation holds at 5% each year, the cost of a kilo of bread will be $1.63
- From this example, it is clear to see "bread costs more" than it did in the past and will cost more in the future
- In other words, "today's dollar buys less" than it did in the past
- The value, or purchasing power, of the dollar decreased through time
- Using the concept of time value of money, various types of debt with different rates or terms can now be compared

From this example it is clear that "bread costs more" now than it did in the past and will cost more in the future. Another way of saying this is that "today's dollar buys less" than it did in the past. This means that the value, or purchasing power, of the dollar decreases through time.

Once the concept of the time-value of money is clear, we will then discuss the various types of debt that a utility can incur to pay for a project.

After we discuss the various types of debt, we can then compare them – no matter whether the types of debt, the rates or the terms are all different – by using the time-value of money.

CHAPTER CONTENTS

CALCULATING CHANGE IN VALUE

Changes in value are calculated by two reciprocal procedures. Calculating value going forward in time is called *compounding*. Calculating value going backward in time is called *discounting*.

Please note that the concepts of "forward" and "backward" are relative. For example, if you know a value in the past, you must compound it to bring it to the present. In a like manner, if you have a value today and want to see what it will be in the future, you must compound. Conversely, if you know a future value (such as a payment you must make in the year 2010) and want to learn its value in today's terms, then you must discount it. And, if you have a value today, and you want to know what that value was in the past, then you also discount.

COMPOUNDING AND DISCOUNTING

To calculate the change in value, you must know the rate of change. Going forward, the rate of change is called the *compound rate*. Going backward, the rate of change is called the *discount rate*.

Calculating Change in Value

Changes in value are calculated by two reciprocal procedures:

- *Compounding*: calculating value going forward in time
- *Discounting*: calculating value going backward in time

THE RATE OF CHANGE

In most cases, both the compound rate and the discount rate are the rate of inflation which can be obtained from the government statistics office or from international financial institutions such as the World Bank or the International Monetary Fund. In certain specific cases, other rates of change can be used, but, in each case, they will be specified.

To compound an amount for a number of years, you must multiply the amount by "the number 1 plus the compound rate (expressed as a decimal)" for the number of years forward that are required. Thus, to get the value of a kilo of bread in two years, if it is worth $1 today and the inflation rate is 5%, you must multiply the $1 times 1.05, twice. ($\$1 \times 1.05 \times 1.05 = \$1.1025$)

To discount an amount for a number of years, you must divide the amount by "the number 1 plus the discount rate (expressed as a decimal)" for the number of years backward that are required. Thus, if a kilo of bread costs $1 today and the inflation rate for the last two years has been 5%, then to determine the price two years ago we must divide the $1 by 1.05, twice. ($\$1/1.05/1.05 = \$0.91$)

Note: Calculations for long-term debt, such as a 40-year bond, can be quite tedious. Calculators make light work of these tasks. Additional details on calculators covered below.

BEWARE!!!

Unfortunately, the economists and the finance specialists have complicated this simple concept. And, even more unfortunately, the calculator manufacturers have followed the direction of the economists and finance specialists. Here is how they have complicated matters:

The Rate of Change

Beware!!!

- Economists and financiers have developed very confusing terminology!
- They call any value a "present value" if it is the earlier of any two values, regardless of whether the value is actually in the past, present or future
- They call any value a "future value" if it is the later of any two values, regardless of whether the value is actually in the past, present or future

Economists and finance specialists refer to the *earlier* value as the "present value" – regardless of when it occurs. They refer to the *later* value as the "future value" – also, regardless of when it occurs. Hand-held calculators use this same convention. Thus, when you want to find out how much a kilo of bread cost ten years ago, and you know the cost today, you enter today's value (which is the later value) as the "future value" on the calculator, and then proceed to calculate the "present value", which is the *earlier* value.

The Rate of Change

- The "present" value is always the earlier value
- If you know the price of bread three years ago, that is the "present value"
- If you want to know what the price is today, after annual inflationary increases of x%, y%, and z%, the price of bread today is the "future value:

The Rate of Change

Remember!!!

- The present value is always the earlier of the two values
- The "future" value is always the later of the two values

Note: Remember, when using a calculator, the "present value" key is always for the *earlier* value and the "future value" key is always for the *later* value.

The Rate of Change

Beware!!!

Remember, when using a calculator, the present value key is always for the earlier value and the future value key is always for the later value.

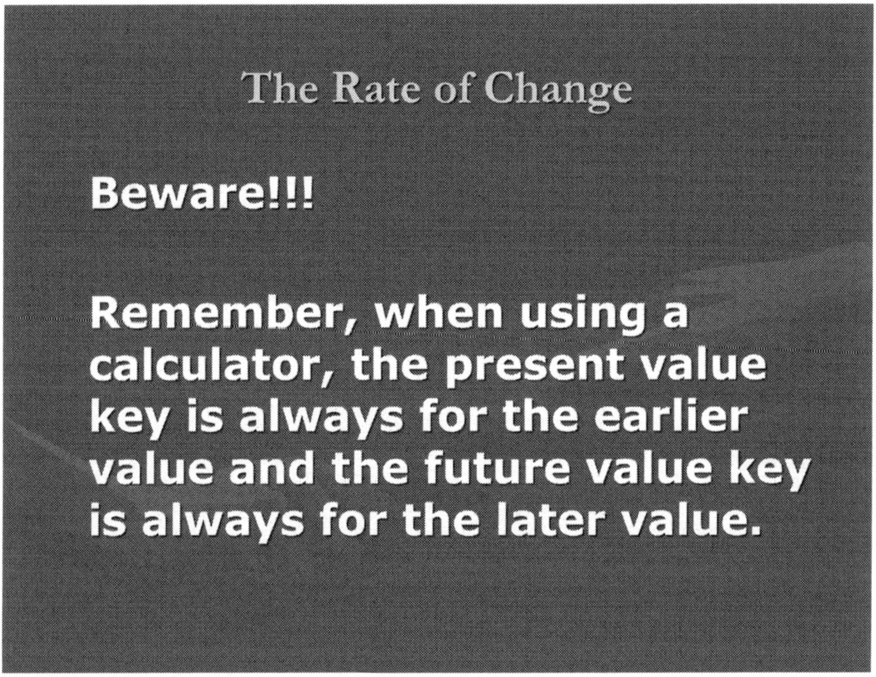

CHOOSING A COMPOUND OR DISCOUNT RATE

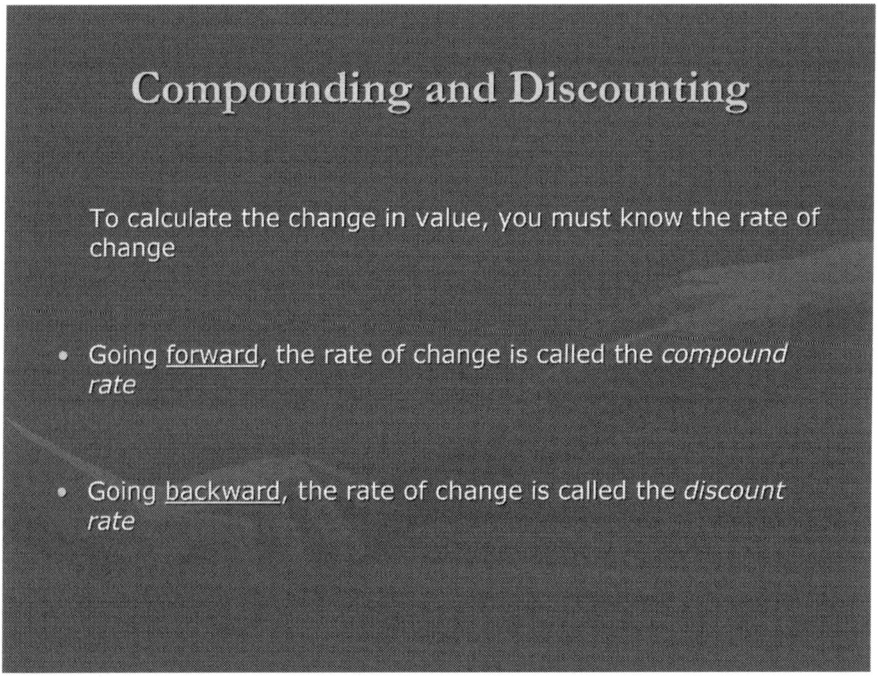

Compounding and Discounting

To calculate the change in value, you must know the rate of change

- Going forward, the rate of change is called the *compound rate*
- Going backward, the rate of change is called the *discount rate*

The Rate of Change

- To **compound** an amount for a number of years, you must multiply the amount by "the number 1 plus the compound rate (expressed as a decimal)" for the number of years forward that are required.

  Compounded Amount = Amount x (1 + compound rate)

- For example, to find the value of a kilo of bread in two years, if it is worth $1 today and the inflation rate is 5%, you must multiply the $1 times 1.05, twice.

  $1 x 1.05 x 1.05 = $1.1025

The Rate of Change

- To **discount** an amount for a number of years, you must divide the amount by "the number 1 plus the discount rate (expressed as a decimal)" for the number of years backward that are required.

  $$\text{Discounted Amount} = \frac{\text{Amount}}{(1 + \text{discount rate})}$$

- For example, if a kilo of bread costs $1 today and the inflation rate for the last two years has been 5%, then to determine the price two years ago you must divide the $1 by 1.05, twice.

  $1/1.05/1.05 = $0.91

There are three rates that are commonly used as compound or discount rates:

- *Inflation rate*: the general rate at which prices increase. This is, by far, the most common. There is usually a government statistics office that publishes historical inflation rates, and a government planning ministry that publishes estimated inflation rates for the next few years.

Choosing a Compound or Discount Rate

Three rates are commonly used as compound or discount rates:

I. **Inflation rate**

- The general rate at which prices increase
- The most commonly used
- Government statistics office publishes historical inflation rates
- Government planning ministry publishes estimated inflation rates for the next few years

- *Reinvestment rate*: the annual rate of return from a known investment. This is used primarily by investment firms to compare a possible investment with a known investment.

Choosing a Compound or Discount Rate

II. **Reinvestment rate**

- The annual rate of return from a known investment
- Used primarily by investment firms to compare a possible investment with a known investment

- *Random rates*: rates chosen by the individual to make specific estimates. For example, someone might want to estimate the cost of an automobile in ten years with the assumption that the price will increase by 5% a year.

Choosing a Compound or Discount Rate

III. **Random rates**

- Rates chosen by the individual to make specific estimates
-
- For example, a person may estimate the cost of an automobile in 10 years with the assumption that the price will increase by 5% a year

Now that we have discussed how the value of money changes over time and have learned how to calculate those changes – by compounding and discounting, we can move on to examine the various types of debt.

TYPES OF DEBT

Types of debt are classified by their principal repayments. To understand the various types of debt, it is useful to construct an Annual Debt Service Payment Schedule for each particular type of loan.

An *Annual Debt Service Payment* is the money paid back for a loan in a specific year. The list, or summary, of all of the Annual Debt Service Payments of a particular loan is called its Annual Debt Service Payment Schedule.

Types of Debt

- Classified by their principal repayments
- Useful to construct an Annual Debt Service Payment Schedule for each particular type of loan
- *Annual Debt Service Payment*: the money paid back for a loan in a specific year
- *Annual Debt Service Payment Schedule*: the list or summary of all the Annual Debt Service Payments of a particular loan

There are three types of loans, or, three methods of repaying the principal of a loan:

- Level payment method
- Level principal method
- Irregular payment method

Types of Debt

There are three types of loans, or, three methods of repaying the principal of a loan:

- **Level payment method**
  - In the U.S., most municipal utility debt is level repayment debt
  - Very suitable for utilities that charge fixed rates (tariffs) to their customers, since the payments are the same each year
- **Level principal method**
  - Widely used in Europe and Asia
  - Since the annual debt service payments decline each year, and since tariffs are fixed, utilities' cash flow increases over time
- **Irregular payment method**
  - Used only when circumstances require it

In the United States, most municipal utility debt is level payment debt. This is very suitable for utilities that have fixed rates (tariffs) that they charge their customers, since the payments are the same each year. In Europe and Asia, the level principal method is used more often. Since the annual debt service payments decline each year with the level principal payment method, and since tariffs are fixed, utilities wind up with considerable extra revenue over time.

The irregular payment method is used only when circumstances require it.

The Level Payment Method

The level payment method means that *the total amount of principal plus interest paid is the same every year*. As the amount of interest declines with every payment that is made, the amount of the principal paid increases by the same amount.

Calculating Annual Debt Service Payments (ADSP) for level payment loans involves the use of a complicated formula, as follows:

$$ADSP = P*(i/(1-(1/(1+i)^n)))$$

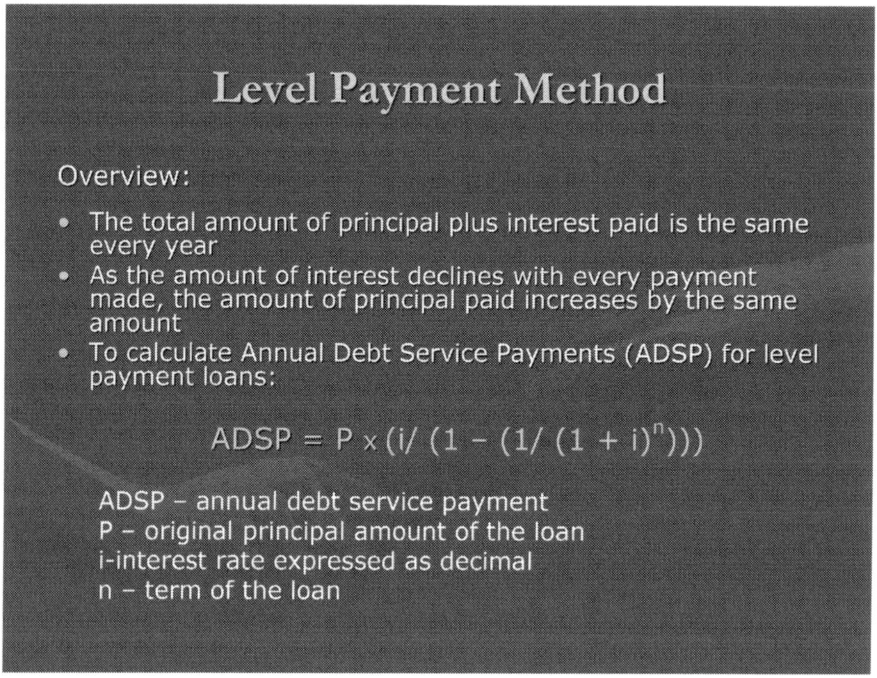

Level Payment Method

Overview:

- The total amount of principal plus interest paid is the same every year
- As the amount of interest declines with every payment made, the amount of principal paid increases by the same amount
- To calculate Annual Debt Service Payments (ADSP) for level payment loans:

$$ADSP = P \times (i/(1-(1/(1+i)^n)))$$

ADSP – annual debt service payment
P – original principal amount of the loan
i-interest rate expressed as decimal
n – term of the loan

Handheld business calculators can deal with this formula very easily with four keystrokes. But for those without such devices, the ADSP on a 40-year loan can take a long time to calculate. Just to illustrate how the formula works, here is the calculation of a two-year loan of $100 at 5% interest.

ADSP – annual debt service payment
P – original principal amount of the loan
i – interest rate expressed as decimal
n – term of the loan

$$ADSP = P \times (i/(1-(1/\ (1+i)^n)))$$

$$ADSP = \$100 \times (.05/(1-(1/(1+.05)2)))$$

$$ADSP = \$100 \times (.05/(1-(1/(1.05)2)))$$

$$ADSP = \$100 \times (.05/(1-(1/1.1025))$$

$$ADSP = \$100 \times (.05/(1-.90702))$$

$$ADSP = \$100 \times (.05/.09298)$$

$$ADSP = \$100 \times .53775$$

$$ADSP = \$53.76$$

Level Payment Method

- Handheld business calculators can compute ADSP easily with four keystrokes
- In absence of a calculator, the following illustrates how the formula works for a 2-year loan of $100 at 5% interest:

Example 2:

$ADSP = P \times (i/\ (1 - (1/\ (1 + i)^n)))$
$ADSP = \$100 \times (.05/(1-(1/(1+.05)2)))$
$ADSP = \$100 \times (.05/(1-(1/(1.05)2)))$
$ADSP = \$100 \times (.05/(1-(1/1.1025))$
$ADSP = \$100 \times (.05/(1-.90702))$
$ADSP = \$100 \times (.05/.09298)$
$ADSP = \$100 \times .53775$
$ADSP = \$53.76$

You can easily prove that this is correct. You know that 5% interest on $100 is $5. So, of the first year's payment of $53.76, you know that $5 of this amount must be interest and the rest, $48.76, must be principal. If you pay off $48.76 of principal in the first year, there is only $51.24 of principal outstanding on the loan at the beginning of the second year. The second year's payment must be $51.24 of principal plus 5% interest. Five percent of $51.24 is $2.52. So, principal of $51.24 and interest of $2.52 equal the ADSP of $53.76.

Level Payment Method

Example 2 (cont.)

ADSP = $53.76

To check the accuracy of your answer:

- 5% interest on $100 is $5
- For the first year's payment of $53.76, you know that $5 of this amount must be interest and the remainder, $48.76, must be principal
- If you pay off $48.76 in the first year, $51.24 is the outstanding principal on the loan at the beginning of the second year
- The second year's payment must be $51.24 of principal plus 5% interest
- 5% of $51.24 is $2.52
- So, principal of $51.24 and interest of $2.52 equal ADSP of $53.76

In going forward, for the purposes of illustrating the various principal payment methods, assume a five year loan of $100,000 with an interest rate of 10%.

As you know, the three variables in any loan transaction are: principal, rate and term. The variables used in this example are:

Principal	P = $100,000
Interest Rate	i = 0.1 (10%)
Term	n = five years

Level Payment Method

Example 3

- In going forward, for the purposes of illustrating the various principal payment methods, assume a 5 year loan of $100,000 with an interest rate of 10%
- Variables used in this example:

Principal	P = $100,000
Interest Rate	i = 0.1 (10%)
Term	n = 5 years

In the case of the given example of a five year $100,000 loan, the Annual Debt Service Payment would look like the following:

$$\text{ADSP} = \$100,000 * \left(0.10/\left(1 - 1/(1.10)^5\right)\right) = \$26,380$$

Year	Prior Balance	Interest	–	Total Annual Payment (ADSP)	=	Principal Payment
1	$100,000	$10,000	–	$26,380	=	$16,380
2	$83,620	$8,362	–	$26,380	=	$18,018
3	$65,603	$6,560	–	$26,380	=	$19,819
4	$45,783	$4,578	–	$26,380	=	$21,801
5	$23,982	$2,398	–	$26,380	=	$23,982
Total		**$31,899**	–	**$131,899**	=	**$100,000**

Level Payment Method

Example 3 (cont.)

In the case of the given example 5 yr. $100,000 loan, the annual debt service payment would look like the following:

$$ADSP = \$100,000 \times (0.10/ (1-1/(1.10)^5))) = \$26,380$$

Table 1

Year	Prior Balance	Interest	-	Total Annual Payment (ADSP)	=	Principal Payment
1	$100,000	$10,000	-	$26,380	=	$16,380
2	$83,620	$8,362	-	$26,380	=	$18,018
3	$65,603	$6,560	-	$26,380	=	$19,819
4	$45,783	$4,578	-	$26,380	=	$21,801
5	$23,982	$2,398	-	$26,380	=	$23,982
Total		**$31,899**	-	**$131,899**	=	**$100,000**

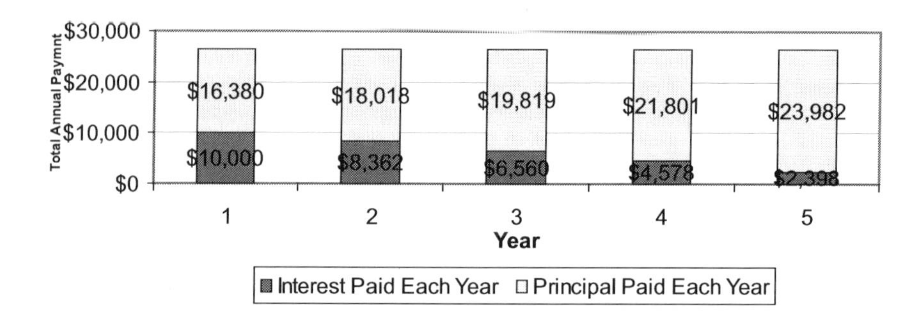

Level Payment Method

Example 3 (cont.)

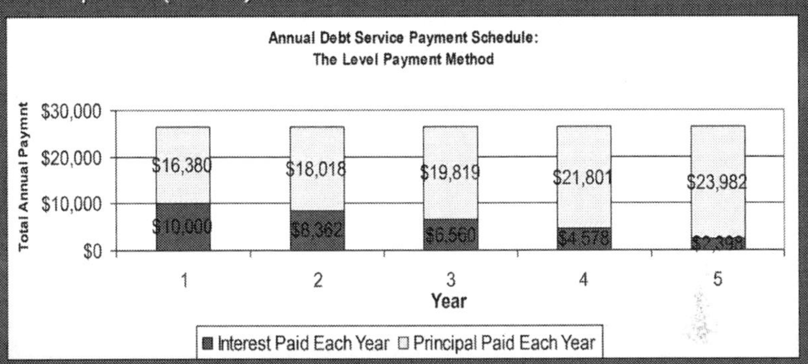

Figure 1: The Annual Debt Service Payment Schedule (also known as an Amortization Schedule) for the Level Payment Method

Creating an Annual Debt Service Payment Schedule for the Level Payment Method:

1. Determine the annual payment (ADSP) using the equation above.
2. Calculate the first year's interest payment. (Interest = i × Principal)
3. Obtain first year's annual principal payment by subtracting the interest fom the ADSP of year one. (Principal = ADSP – interest)
4. Subtract the year one principal payment from the original principal amount.
5. Calculate the second year's interest payment (Interest = i × Outstanding Principal Balance)
6. Repeat the above process for each year of the loan term.

Level Payment Method

Creating an Annual Debt Service Payment Schedule for the Level Payment Method:

1. Determine the annual payment (ADSP) using the equation above
2. Calculate the first year's interest payment (Interest = i x Principal)
3. Obtain first year's annual principal payment by subtracting the interest from the ADSP of year one (Principal = ADSP – interest)
4. Subtract the year one principal payment from the original principal amount
5. Calculate the second year's interest payment (Interest = i x Outstanding Principal Balance)
6. Repeat the above process for each year of the loan term

<u>Summary of the Level Payment Method:</u>

- The annual payments are always the same.
- The amount of principal paid each year increases by the exact amount in which the interest payment decreases.
- The total of all of the annual principal payments equals the original principal amount of the loan.

Level Payment Method

Summary of the Level Payment Method:

- The annual payments are always the same
- The amount of principal paid each year increases by the exact amount in which the interest payment decreases
- The total of all of the annual principal payments equals the original principal amount of the loan

The Level Principal Payment Method

The Level Principal Payment Method means that the total amount of principal is the same from year to year and the interest payment is simply calculated using the remaining balance. As the amount of interest declines with every payment that is made, the total Annual Debt Service Payment also decreases.

<u>Calculating Annual Debt Service Payments for Loans using the Level Principal Payment Method:</u>

P – original principal amount of the loan
n – term of the loan

Notice that the Annual Principal Payment is the original principal amount divided by the term (P/n).

Level Principal Payment Method

Overview:

- The total amount of principal repaid is the same from year to year and the interest payment is simply calculated using the remaining balance
- As the amount of interest declines with every payment that is made, the total annual debt service payment also decreases
- Calculating ADSP for loans using the Level Principal Payment Method:
  P = original principal amount of the loan
  n = term of the loan
- The annual principal payment is the original principal amount divided by the term (P/n)

In the case of the example of a five year loan of $100,000 at 10% interest, the Annual Debt Service Payment would look like the following:

$$\text{Annual principal payment} = \$100,000/5 = \$20,000$$

Year	Prior Balance	Interest	+	Principal Payment	=	Total Annual Payment (ADSP)
1	$100,000	$10,000	+	$20,000	=	$30,000
2	$80,000	$8,000	+	$20,000	=	$28,000
3	$60,000	$6,000	+	$20,000	=	$26,000
4	$40,000	$4,000	+	$20,000	=	$24,000
5	$20,000	$2,000	+	$20,000	=	$22,000
Total		**$30,000**	**+**	**$100,000**	**=**	**$130,000**

Level Principal Payment Method

Example 4

In the case of a 5 yr. loan of $100,000 at 10% interest, the annual debt service payment would look like the following:

Annual principal payment = $100,000 / 5 = $20,000

Table 2

Year	Prior Balance	Interest	+	Principal Payment	=	Total Annual Payment (ADSP)
1	$100,000	$10,000	+	$20,000	=	$30,000
2	$80,000	$8,000	+	$20,000	=	$28,000
3	$60,000	$6,000	+	$20,000	=	$26,000
4	$40,000	$4,000	+	$20,000	=	$24,000
5	$20,000	$2,000	+	$20,000	=	$22,000
Total		**$30,000**	**+**	**$100,000**	**=**	**$130,000**

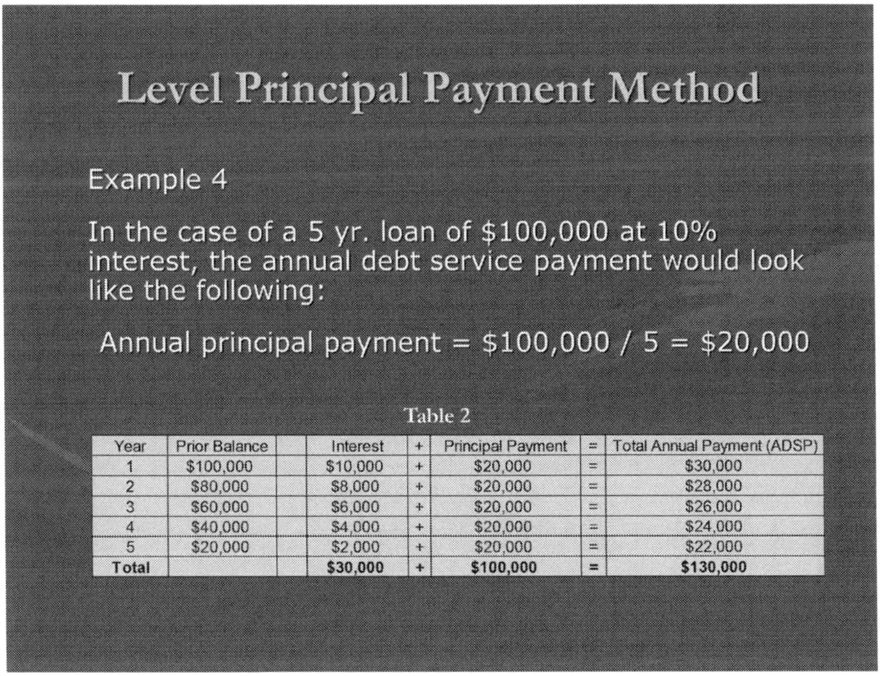

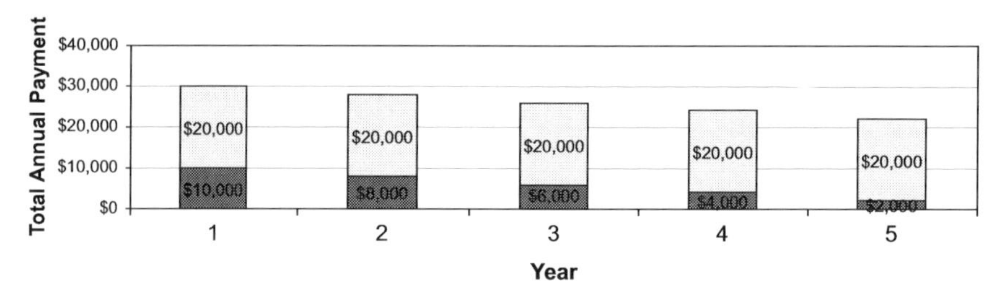

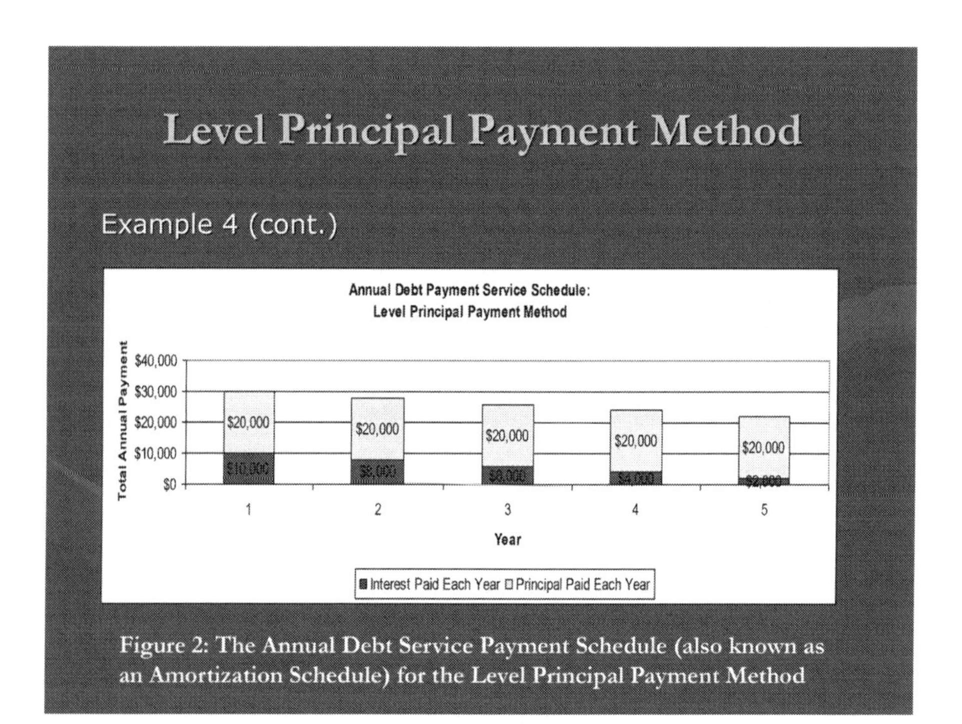

Figure 2: The Annual Debt Service Payment Schedule (also known as an Amortization Schedule) for the Level Principal Payment Method

Creating an Annual Debt Service Payment Schedule for the Level Principal Payment Method:

1. Calculate the level Annual Principal Payment by dividing the original principal amount by the number of years in the term of the loan. (Annual Principal Payment = P/n)
2. Calculate the first year's interest payment by multiplying the original principal amount of the loan by the rate of interest. (Interest = $i * P$)
3. Obtain the first year's annual payment by adding the level principal payment to the first year's annual interest payment.
(1st year ADSP = interest + principal payment)
4. Obtain the outstanding principal balance at the end of the first year by subtracting the first year's level principal payment from the original principal amount of the loan.
(Outstanding Balance = P – ADSP)
5. Calculate the next year's annual interest payment by multiplying the outstanding principal balance from the preceding year by the rate of interest.
(2nd year interest = outstanding balance*i)
6. Calculate the next year's annual payment by adding that year's annual interest payment to the level principal payment.
(2nd year ADSP = 2nd year interest + principal payment)
7. Continue the process for each year of the term.

Level Principal Payment Method

Creating an Annual Debt Service Payment Schedule for the Level Principal Payment Method:

1. Calculate the annual principal payment by dividing the original principal amount by the number of years in the term of the loan (Annual principal payment = P/n)
2. Calculate the first year's interest payment by multiplying the original principal amount of the loan by the rate of interest (Interest = i * P)
3. Obtain the first year's annual payment by adding the level principal payment to the first year's annual interest payment (1st year ADSP = interest + principal payment)
4. Obtain the outstanding principal balance at the end of the first year by subtracting the first year's level principal payment from the original principal amount of the loan (Outstanding Balance = P – ADSP)
5. Calculate the second year's interest payment by multiplying the original principal amount of the loan by the rate of interest (Interest = i * P)
6. Calculate the next year's annual payment by adding that year's annual interest payment to the level principal payment (2nd year ADSP = 2nd year interest + principal payment)
7. Continue the process for each year of the term

Summary of the Level Principal Payment Method:

- The annual principal payments are the same each year.
- Both the annual interest payment and the annual payment decrease each year.
- The total of all of the annual principal payments, equals to the original principal amount of the loan.
- Note that if there were no interest (an interest rate = 0%) then the level principal method would be the same at the level payment method, i.e., the payments would be the same each year.

Level Principal Payment Method

Summary of the Level Principal Payment Method:

- The annual principal payments are the same each year
- Both the annual interest payment and the annual payment itself decrease each year
- The total of all of the annual principal payments, equals to the original principal amount of the loan
- Note, if there were no interest (interest rate = 0%) then the level principal method would be the same at the level payment method, i.e., the payments would be the same each year

The Irregular Payment Method

Unlike both the Level Payment Method and the Level Principal Payment Method, which have at least one constant, the irregular method of debt service payment has none. There are very few scenarios where the irregular method is used. Some examples of such circumstances are:

Irregular Payment Method

Overview:

- The Irregular Payment Method has no constants
- There are very few circumstances where the irregular method is used
- Some example of such circumstances are:
  - Delayed Benefits
  - Expectation of Future Income
  - Favorable Interest Rates

Scenario #1: Delayed Benefits

A utility may issue a bond or obtain a loan prior to the construction of its project. In such case the project may require two or more years to complete, and the utility may not want to begin charging its users until the benefits of the project are available to them. Thus, if the goal is to minimize the cost to the ratepayers prior to the actual operation of the project, then, the utility will elect not to pay principal or interest payments in the first three years. Financially strong utilities will be offered an opportunity to defer such payments. To do that, a utility will issue a bond or take out a loan and the lender or bondholders will still receive interest payments when due. But the utility would make such payments from borrowed funds. In short, the utility over borrows the precise amount it needs to make up for the missing user fees. In such case, there are no principal repayments in the early years. However, once the project is complete and the new user fees begin to be paid, the tenor of the bond or loan generally reverts to the level payment method or the level principal payment method. Thus, in the case of a delayed benefit, the principal payments are irregular, but only in the beginning years of the loan.

Irregular Payment Method

Scenario #1: Delayed Benefits

- A utility may issue or obtain a loan prior to the construction of its project
- In such case, the project may require two or more years to complete
- The utility may not want to begin charging its users until the benefits of the project are available to them
- If the goal is to minimize the cost to the ratepayers prior to the actual operation of the project, then, the utility will elect not to pay principal or interest payments in the first three years
- Financially strong utilities will be offered an opportunity to defer such payments

Irregular Payment Method

Scenario #1: Delayed Benefits (cont.)

- To do that, a utility will issue a bond or take out a loan and the lender or bondholders will still receive interest payments when due, but the utility will make such payments from borrowed funds
- In short, the utility over-borrows the additional amount it needs to make the first few annual debt service payments
- In such case, there are no principal repayments in the early years
- However, once the project is complete and the new user fees begin to be paid, the tenor of the bond or loan generally reverts to the level payment method or the level principal payment method
- Thus, in the case of a delayed benefit, the principal payments are irregular, but only in the beginning years of the loan

Scenario #2: Expectation of Future Income

Another circumstance which might warrant the use of the irregular method is when the utility has a realistic expectation of some non-recurring future income. For example, a system may be planning to sell some of its property. As such, it should certainly be able to estimate the amount of gain from the sale and plan to set aside a portion of that gain to pay off the loan principal. This would warrant using the irregular payment method with principal payments skewed to those years immediately after the planned sale date of property. In other words, once the property was sold, the utility would use those funds to pay off part of its loan. This would make the principal payment schedule irregular.

Irregular Payment Method

Scenario #2: Expectation of Future Income

- This Circumstance may occur when the utility has a realistic expectation of some non-recurring future income
- For example, a system may be planning to sell some of its property
- The amount of gain from the sale should be estimated and a portion of the gain set aside to pay off the loan principal
- Principal payments should be skewed to those years immediately after the planned sale date of the property

Scenario #3: Favorable Interest Rates

Another circumstance which might also warrant the use of an irregular payment method, arises not from within the system, but from without the system. It actually arises from within the credit market, and is especially true in the case of a bond market. For example, there may be a large number of buyers willing to buy bonds of certain maturities (e.g., ten years) but not others (e.g., five years). The interest rates on the ten-year maturities should be favorable to the utility that is borrowing. A utility can set its annual debt service schedule with principal payments skewed to those favorable maturities to take advantage of the favorable rates.

Irregular Payment Method

Scenario #3: Favorable Interest Rates

- This circumstance arises from within the credit market, and is especially true in the case of the bond market
- For example, there may be a large number of buyers willing to buy bonds of certain maturities (e.g., ten years) but not others (e.g., five years)
- Interest rates on the ten year maturities should be favorable to the utility that is borrowing
-
- A utility can set its annual debt service schedule with principal payments skewed to those favorable maturities to take advantage of the favorable rates

Creating Annual Debt Service Payment Schedules Using the Irregular Method:

1. Since the irregular payment method, in and of itself, is a list of principal payments on certain maturities, all that remains to be done is to calculate the interest payable each year on the outstanding balance.
2. Thereinafter, all that needs be done is to subtract that year's principal payment from the outstanding balance, and repeat the above procedure.

Irregular Payment Method

Creating Annual Debt Service Payment Schedules Using the Irregular Method:

- Since the irregular payment method, in and of itself, is a list of principal payments on certain maturities, all that remains to be done is to calculate the interest payable each year on the outstanding balance
- Thereinafter, all that needs to be done is to subtract that year's principal payment from the outstanding balance, and repeat the above procedure

COMPARING TYPES OF DEBT

Comparing Loans

- The most important method for comparing financing options is the *Present Value Method*
- Present Value Method:
  - Goal: To determine the present value for all project costs under financing options so that they can be compared
  - Discounting will be used
  - Note: Final values depend upon the discount factor or range of discount factors chosen for the analysis
  - Only loans with the <u>same original principal amount</u> may be compared

There are a number of methods for comparing financing options. Probably the most important one is the Present Value Method.

PRESENT VALUE METHOD

The goal of the Present Value Method is to determine the present value for all project costs under various financing options so that they can be compared. This method uses discounting, which was discussed above. Please note that, as with any calculation involving discounting, the final values depend upon the discount factor or range of discount factors chosen for the analysis. This method can only be used to compare *loans with the same original principal amount*. Thus a loan of $10,000 can not be compared to a loan of $12,000.

To illustrate the method, assume a utility manager is facing a 4% – 6% rate of inflation and has the following options, taken from a yield curve, for a $100,000 loan:

Option A: 5% with a five year term (no points/fees)
Option B: 6% with a ten year term (no points/fees)

Present Value Method

Example 5

Assume a utility manager is facing a 4% - 6% rate of inflation and has the following options, taken from a yield curve, for a $100,000 loan:

Option A: 5% with a 5 year term (no points/fees)

Option B: 6% with a 10 year term (no points/fees)

Following are tables of amortization schedules (or annual debt service payment schedules) for the two hypothetical loans using the level payment method.

Option A

Calculating the Annual Debt Service Payment:

$$\text{ADSP} = \$100,000*(0.05/(1 - 1/(1.05)^5) = \$23,097$$

Year	Prior Balance	Interest	Annual Payment (ADSP)	Principal Payment
1	100,000	5,000	23,097	18,097
2	81,903	4,095	23,097	19,002
3	62,900	3,145	23,097	19,952
4	42,948	2,147	23,097	20,950
5	21,998	1,100	23,097	21,998
Total (USD)		**15,487**	**115,487**	**100,000**

Option B

Calculating the Annual Debt Service Payment:

$$\text{ADSP} = \$100,000*\left(0.06/\left(1 - 1/(1.06)^{10}\right)\right) = \$13,587$$

Year	Prior Balance	Interest	Annual Payment (ADSP)	Principal Payment
1	100,000	6,000	13,587	7,587
2	92,413	5,545	13,587	8,042
3	84,371	13,587	13,587	8,525
4	75,847	13,587	13,587	9,036
5	66,811	13,587	13,587	9,578
6	57,233	13,587	13,587	10,153
7	47,080	13,587	13,587	10,762
8	36,318	13,587	13,587	11,408
9	24,910	13,587	13,587	12,092
10	12,818	13,587	13,587	12,818
Total (USD)		**62,745**	**135,868**	**100,000**

Present Value Method

Example 5 (cont.)

Below are tables of amortization schedules (or annual debt service payment schedules) for the two hypothetical loans using the level payment method:

Option A

Yr	Prior Balance	Interest	Annual Payment (ADSP)	Principal Payment
1	100,000	5,000	23,097	18,097
2	81,903	4,095	23,097	19,002
3	62,900	3,145	23,097	19,952
4	42,948	2,147	23,097	20,950
5	21,998	1,100	23,097	21,998
Total (USD)		15,487	115,487	100,000

Option B

Yr	Prior Balance	Interest	Annual Payment (ADSP)	Principal Payment
1	100,000	6,000	13,587	7,587
2	92,413	5,545	13,587	8,042
3	84,371	13,587	13,587	8,525
4	75,847	13,587	13,587	9,036
5	66,811	13,587	13,587	9,578
6	57,233	13,587	13,587	10,153
7	47,080	13,587	13,587	10,762
8	36,318	13,587	13,587	11,408
9	24,910	13,587	13,587	12,092
10	12,818	13,587	13,587	12,818
Total (USD)		62,745	135,868	100,000

Option A: ADSP = $100,000 x (0.05/(1-1/(1.05)^5))) = **$23,097**

Option B: ADSP = $100,000 x (0.06/1-1/(1.06)^{10}))) = **$13,587**

If one wanted to compare the total cost of Options A and B, it is tempting to simply add up all the Annual Debt Service Payments as shown above and compare $135,868 to $115,487. THIS IS NOT CORRECT FINANCIAL ANALYSIS. The dollars expressed in the Annual Debt Service Payment schedules are at different times and must be discounted to a present value for comparison. The correct question is “How much is total cost of each loan in today’s dollars (present value)?” For each Annual Debt Service Payment, convert the future value to present value through discounting, and sum the present values to get the total present value of the loan.

$$\text{Discount factor} = 1/(1 + \%\ \text{discountrate})$$

$$\text{Present Value} = \text{Future Value} * \left(\text{discount factor}^{\text{years of difference}}\right)$$

Present Value Method

Example 5 (cont.)

- When comparing the total cost of Options A and B, do not simply add up all the annual debt service payments as shown on the previous slide and compare $135,868 to $115,487 – **This is not correct financial analysis!**

- The dollars expressed in the annual debt service payment schedules are at different times and must be discounted to a present value for comparison
- You must determine the total cost of each loan in today's dollars (present value)
- For each annual debt service payment, convert the future value to present value through discounting, and sum the present values to get the total present value of the loan

Discount factor = 1/ (1 + % Discount Rate)

Present Value = Future Value x (Discount Factor $^{\text{years of difference}}$)

For the two options presented, the 4% inflation rate will first be used as the discount rate. Thus, the discount factor for this example is 1/1.04 and the present values of the Annual Debt Service Payments are as follows:

Option A: Present Value of Annual Debt Service Payments (ADSP)

Year	Annual Payment (ADSP)	(4% discount rate)	Present Value of ADSP
1	23,097	X $(1/1.04^1)$=	22,209
2	23,097	X $(1/1.04^2)$=	21,355
3	23,097	X $(1/1.04^3)$=	20,534
4	23,097	X $(1/1.04^4)$=	19,744
5	23,097	X $(1/1.04^5)$=	18,984
Total	**[115,487]**		**102,826**

Option B: Present Value of Annual Debt Service Payments (ADSP)

Year	Annual Payment (ADSP)	(4% discount rate)	Present Value of ADSP
1	13,587	X $(1/1.04^1)$=	13,064
2	13,587	X $(1/1.04^2)$=	12,562
3	13,587	X $(1/1.04^3)$=	12,079
4	13,587	X $(1/1.04^4)$=	11,614
5	13,587	X $(1/1.04^5)$=	11,167
6	13,587	X $(1/1.04^6)$=	10,738
7	13,587	X $(1/1.04^7)$=	10,325
8	13,587	X $(1/1.04^8)$=	9,928
9	13,587	X $(1/1.04^9)$=	9,546
10	13,587	x $(1/1.04^{10})$=	9,179
Total	**[135,868]**		**110,201**

Present Value Method

Example 5 (cont.)

For the two options presented, the 4% inflation rate will first be used as the discount rate. Thus, the discount factor for this example is 1/1.04 and the present values of the annual debt service payments are as follows:

Option A: Present Value of Annual Debt Service Payment (ADSP)

Yr	Annual Payment (ADSP)	(4% discount rate)	Present Value of ADSP
1	23,097	X $(1/1.04^1)$ =	22,209
2	23,097	X $(1/1.04^2)$ =	21,355
3	23,097	X $(1/1.04^3)$ =	20,534
4	23,097	X $(1/1.04^4)$ =	19,744
5	23,097	X $(1/1.04^5)$ =	18,984
Total	[115,487]		102,826

Option B: Present Value of Annual Debt Service Payment (ADSP)

Yr	Annual Payment (ADSP)	(4% discount rate)	Present Value of ADSP
1	13,587	X $(1/1.04^1)$ =	13,064
2	13,587	X $(1/1.04^2)$ =	12,562
3	13,587	X $(1/1.04^3)$ =	12,079
4	13,587	X $(1/1.04^4)$ =	11,614
5	13,587	X $(1/1.04^5)$ =	11,167
6	13,587	X $(1/1.04^6)$ =	10,738
7	13,587	X $(1/1.04^7)$ =	10,325
8	13,587	X $(1/1.04^8)$ =	9,928
9	13,587	X $(1/1.04^9)$ =	9,546
10	13,587	X $(1/1.04^{10})$ =	9,179
Total	[135,868]		110,201

Now that all the future values have been converted to present values and summed, the total values can be compared. Option B turns out to be more expensive than Option A, but by significantly less than if one attempted to compare them without first converting to present value.

Present Value Method

Example 5 (cont.)

Now that all the future values have been converted to present values and summed, the total values can be compared. Option B turns out to be more expensive than Option A, but by significantly less than if one attempted to compare them without first converting to present value.

However, the rate of inflation could be as high as 6% so a second analysis must be done with this discount rate. Now the discount factor is 1/1.06 and the present values of the annual debt service payments are as follows:

Option A: Present Value of Annual Debt Service Payments (ADSP)

Year	Annual Payment (ADSP)	(6% discount rate)	Present Value of ADSP
1	23,097	x $(1/1.06^1)$=	21,790
2	23,097	x $(1/1.06^2)$=	20,557
3	23,097	x $(1/1.06^3)$=	19,393
4	23,097	x $(1/1.06^4)$=	18,295
5	23,097	x $(1/1.06^5)$=	17,260
Total	**[115,487]**		**97,295**

Option B: Present Value of Annual Debt Service Payments (ADSP)

Year	Annual Payment (ADSP)	(6% discount rate)	Present Value of ADSP
1	13,587	x $(1/1.06^1)$=	12,818
2	13,587	x $(1/1.06^2)$=	12,092
3	13,587	x $(1/1.06^3)$=	11,408
4	13,587	x $(1/1.06^4)$=	10,762
5	13,587	x $(1/1.06^5)$=	10,153
6	13,587	x $(1/1.06^6)$=	9,578
7	13,587	x $(1/1.06^7)$=	9,036
8	13,587	x $(1/1.06^8)$=	8,525
9	13,587	x $(1/1.06^9)$=	8,042
10	13,587	x $(1/1.06^{10})$=	7,587
Total	**[135,868]**		**100,000**

Present Value Method

Example 5 (cont.)

However, the rate of inflation could be as high as 6% so a second analysis must be completed with this discount rate. Now the discount factor is 1/1.06 and the present values of the annual debt service payments are as follows:

Option A : Present Value of Annual Debt Service Payment (ADSP)

Yr	Annual Payment (ADSP)	(4% discount rate)	Present Value of ADSP
1	23,097	X $(1/1.06^1)$ =	21,790
2	23,097	X $(1/1.06^2)$ =	20,557
3	23,097	X $(1/1.06^3)$ =	19,393
4	23,097	X $(1/1.06^4)$ =	18,295
5	23,097	X $(1/1.06^5)$ =	17,260
Total	[115,487]		97,295

Option B : Present Value of Annual Debt Service Payment (ADSP)

Yr	Annual Payment (ADSP)	(4% discount rate)	Present Value of ADSP
1	13,587	X $(1/1.06^1)$ =	12,818
2	13,587	X $(1/1.06^2)$ =	12,092
3	13,587	X $(1/1.06^3)$ =	11,408
4	13,587	X $(1/1.06^4)$ =	10,762
5	13,587	X $(1/1.06^5)$ =	10,153
6	13,587	X $(1/1.06^6)$ =	9,578
7	13,587	X $(1/1.06^7)$ =	9,036
8	13,587	X $(1/1.06^8)$ =	8,525
9	13,587	X $(1/1.06^9)$ =	8,042
10	13,587	X $(1/1.06^{10})$ =	7,587
Total	[135,868]		100,000

With a higher discount rate, the present value of the loan decreases. This is a highly unlikely situation because the discount rate is above or equal to the interest rate in both cases. The present value of Option A is less than the original principal of $100,000, so the borrower is making money while the lender is loosing money, assuming inflation is 6%. In Option B, the interest rate is exactly equal to the discount rate so the total present value of the loan is equal to the principal of the loan. Note that the actual dollar values paid have not changed for either option; the only difference is that they have been analyzed with a higher, plausible discount rate.

Present Value Method

Example 5 (cont.)

- With a higher discount rate, the present value of the loan decreases
- This is a highly unlikely situation because the discount rate is above or equal to the interest rate in both cases
- The present value of Option A is less than the original principal of $100,000 so the borrower is making money while the lender is loosing money, assuming inflation is 6%
- In Option B, the interest rate is exactly equal to the discount rate so the total present value of the loan is equal to the principal of the loan
- The actual dollar values paid have not changed for either option
- The only difference is they have been analyzed with a higher, plausible discount rate

Method Rules:

1. Write out the Annual Debt Service Payment schedule (also known as the amortization schedule) for each financing option, including the cash amount of each annual payment and years from now each will need to be paid.
2. Decide on a range of appropriate discount rates.
3. Convert each annual payment to its present value using the lowest possible discount factor.
4. Sum up the present values per financing option using the lowest discount factor.
5. Compare the financing options for using the lowest discount factor.
6. Repeat steps 2–3 for the highest possible discount factor.

Present Value Method

Method Rules:

1. Write out the annual debt service payment schedule (also known as the amoritization schedule) for each financing option, including the cash amount of each annual payment and years from now each will need to be paid
2. Decide on a range of appropriate discount rates
3. Convert each annual payment to its present value using the lowest possible discount factor
4. Sum up the present value per financing option using the lowest discount factor
5. Compare the financing options for using the lowest discount factor
6. Repeat steps 2 - 3 for the highest possible discount factor

Present Value Method with Fees

Loan fees arise for a variety of reasons, such as engineers' fees, lawyers' fees or bank fees (called "points"). They can be sorted into three categories and taken into account as follows:

Types of Fees	Treatment
One-time only fees (such as points)	Add to project cost
Annual fees (flat dollar amounts, independent of principle)	Add to annual principal and interest payments
Based on % of outstanding principal balance	Add to the interest rate

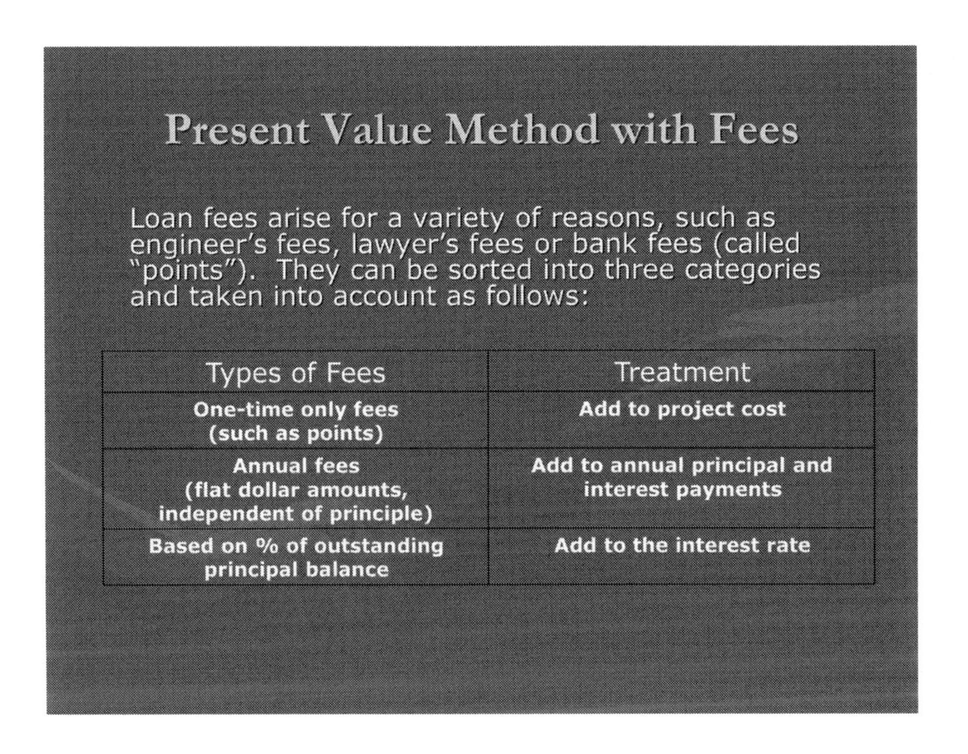

FACTORS THAT INFLATE PROJECT COSTS

Four Factors that Inflate the Cost of Projects

There are four major factors that directly or indirectly increase the cost of funding a project:

I) *Financing costs* – expenses directly related to the obtaining of funds, including costs which are not needed for the project but required by a particular financial method.
II) *Delay* – the associated costs, either as financing costs or through the effects of the time value of money, from a delayed start of the project.
III) *Ineligibility* – the portion of the total project cost which can not be financed through a particular financing program.
IV) *Coverage* – the amount the ratepayers' annual payments must exceed the annual debt service due.

Factors that Inflate the Cost of Projects

There are four major factors that directly or indirectly increase the cost of funding a project:

I. **Financing costs** – expenses directly related to the obtaining of funds, including costs which are not needed for the project but required by a particular financial method
II. **Delay** – the associated costs, either as financing costs or through the effects of the time value of money, from a delayed start of the project
III. **Ineligibility** – the portion of the total project cost which can not be financed through a particular financing program
IV. **Coverage** – the amount the ratepayer's annual payments must exceed the annual debt service due

Financing Costs

These are the costs of obtaining the funds themselves or of making the funding possible in the first place. These are not abstract concepts like term and rate, but fees that must be considered in the overall project cost. Calculate them into the project cost by adding them to the principal. Below are the eight most common forms of financing costs:

- *Commitment fees or points* – an additional payment to the lender that is paid upon acceptance of the bank's commitment to lend.
- *Financial advisory fees* – fees paid to financial advisors hired by the utility, independent of the lender, who advise the utility at every step of the negotiations.
- *Counsel fees* – fees paid to lawyers hired by the utility, independent of the lender, who assist in negotiations.
- *Servicing fees* – administrative costs of collecting loan or bond payments.
- *Placement fees* – (bond transactions only) compensation to the investment banker for arranging the sale of bonds.

I. Financing Costs

The following are the most common forms of financing costs associated with obtaining the funds or of making the funding possible in the first place. These fees must be calculated into the overall project cost.

- *Commitment fees or points* – additional payment to the lender paid upon acceptance of the bank's commitment to lend
- *Financial advisory fees* – fees paid to financial advisors hired by the utility, independent of the lender, who advise the utility at every step of the negotiations
- *Counsel fees* – fees paid to lawyers hired by the utility, independent of the lender, who assist in negotiations
- *Servicing fees* – administrative costs of collecting loan or bond payments
- Placement fees – (bond transaction only) compensation to the investment banker for arranging the sale of bonds

- *Credit enhancement fees* – (bond transactions only) when an eligible utility elects to enhance their credit through the use of municipal bond insurance or bank letters of credit, there are associated fees.
- *Rating agency fees* – (bond transactions only) when a utility issues a bond, its credit is independently evaluated by a rating agency which charges these fees.
- *Printing and miscellaneous costs* – (bond transactions only) costs to various third parties which are generally easily identifiable in advance and do not involve large sums of money. These might include surveys, appraisals, and cost of printing bond documents.

I. Financing Costs (Cont.)

- *Placement fees* – (bond transaction only) compensation to the investment banker for arranging the sale of bonds
- *Credit enhancement fees* – (bond transactions only) when an eligible utility elects to enhance their credit through the use of municipal bond insurance or bank letters of credit, there are associated fees
- *Rating agency fees* – (bond transactions only) when a utility issues a bond, its credit is independently evaluated by a rating agency which charges these fees
- *Printing and miscellaneous costs* – (bond transactions only) costs to various third parties which are generally easily identifiable in advance and do not involve large sums of money (e.g., surveys, appraisals, cost of printing bond documents, etc.)

Other imposed costs may include other insurance, appraisals, and audits which may be required by government regulation in order to proceed with the project and the project financing.

Delay

Budgets must take into account inflation and the time value of money, as was outlined in the beginning of this chapter. For example, at the beginning of "Year 1" an infrastructure project is estimated to cost $100,000. The utility decides to implement the project, but is forced to wait three years from bureaucratic delays. In three years the "cost" is no longer $100,000, the project cost must be compounded to a future value. Assuming a 4% rate of inflation the new project cost will be over $12,000 more than the original estimate: future value of project cost = $\$100{,}000 * (1.04)^3 = \$112{,}486$

II. Delay

- Budgets must take into account inflation and the time value of money
- For example:
  - At the beginning of "Year 1" an infrastructure project is estimated to cost $100,000
  - The utility decides to implement the project, but is forced to wait 3 years from bureaucratic delays
  - In three years the cost is no longer $100,000, the project cost must be compounded to a future value
  - Assuming a 4% rate of inflation, the new project cost will be over $12,000 more than the original estimate: future value of project cost = $\$100{,}000 \times (1.04)^3 = \$112{,}486$

Ineligibility

Ineligibility means that a lender's particular program to fund the system's project will not permit certain categories of costs to be included. Ineligible costs do not directly add to project costs, but create a second project, from a financial perspective, and must be financed in other ways.

- *Quantitative ineligibility* – refers to a percentage of the total cost which a lender will not finance. Some lenders will only finance 50%, or 75%, of project costs.
- *Qualitative ineligibility* – refers to whole categories of costs which will not be considered for financing. An example is the purchase of land.

III. Ineligibility

A lender's particular program to fund the system's project will not permit certain categories of costs to be included. Ineligible costs do not directly add to project costs, but create a second project, from a financial perspective, and must be financed in other ways:

- *Quantitative ineligibility* – a percentage of the total cost which a lender will not finance, e.g., some lenders will only finance 50%, or 75% of project costs
- *Qualitative ineligibility* – whole categories of costs which will not be considered for financing, e.g., the purchase of land

Coverage

Coverage is the amount – as required by the lender – by which a utility's net income must exceed its annual debt service payments.

Coverage ratio – the ratio of Cash Available for Debt Service to the Annual Debt Service Payment.

For example if the cash available for debt service was $115 and the annual debt service payment was $100 then there would be 115% coverage or a 1.15 coverage ratio.

Many lenders require a certain debt service coverage ratio to minimize risk associated with the loan. This coverage is not collected by the lender; but the lender requires it to be there. Thus, this coverage constitutes mandated excess funds which can be used for any lawful purpose, such as saving for a reserve fund or used to prepay loans, after the annual debt service payment is made.

IV. Coverage

- *Coverage* – amount, as required by the lender, by which a utility's net income must exceed its annual debt service payments
- *Coverage ratio* – ratio of Cash Available for Debt Service to the Annual Debt Service Payment
- For example, if the cash available for debt service was $115 and the annual debt service payment was $100 then there would be 115% coverage or a 1.15 coverage ratio
- Many lenders require a certain debt service coverage ratio to minimize risk associated with the loan
- Coverage is not collected by the lender; but the lender requires it to be there
- This coverage constitutes mandated excess funds which can be used for any lawful purpose, such as saving for a reserve fund or used to repay loans, after the annual debt service payment is made

Questions and Answers

Question #1:

The cost of a candy bar five years ago was $0.50 and the inflation rate has been 4% each year for the last five years. Furthermore, the future rate of inflation will be 5% each year for the following three years.

Part A: What is the cost of a candy bar today?

Part B: What will be the cost of a candy bar three years from now?

Question #2:

If a can of soda costs $1.00 today and the inflation rate for the last four years has been 3%, what was the price of the can of soda four years ago?

Question #3:

Calculate the Annual Debt Service Payment (ADSP) using the Level Payment Method for a five year loan of $1000 at 8% interest.

Question #4:

Calculate the Annual Debt Service Payment (ADSP) using the Level Payment Method for a three year loan of $500 at 4% interest.

Question #5:

Calculate the annual principal payment of a five year loan of $175,000 at 10% interest using the Level Principal Payment Method.

Question #6:

Calculate the annual principal payment of a ten year loan of $250,000 at 8% interest using the Level Principal Payment Method.

Question #7:

XYZ Company would like to purchase machinery and equipment for their wastewater system. The bond is for $100,000 and their financial advisor recommends paying off 30%, or $30,000, at the end of the fifth year and the balance, 70% or $70,000, at the end of the eighth year. The financial advisor speculates the five-year maturity will carry an Interest rate of 5% and the eight-year maturity will carry an Interest rate of 7%. Using the Irregular Principal Payment Method, create an annual debt service payment schedule for the whole bond.

Question #8:

Of the following short term loans, please create a table for each and compare. Using the Level Payment Method, which loan would be least costly to the borrower?

Loan A: four year loan of $50,000 at 10% Interest

Loan B: six year loan of $50,000 at 8% Interest

Answer #1:

Part A:

The price of the candy bar increased at a rate of 4% each year for the past five years.
Therefore, the cost of a candy bar today will be:

$$\$0.50 \times 1.04 \times 1.04 \times 1.04 \times 1.04 \times 1.04 = \$0.6083264512 \text{ or } \$0.61$$

Year 1 (four years ago), cost of candy bar = $\$0.50 \times (1 + 0.04) = \$0.52$

Year 2 (three years ago), cost of candy bar = $\$0.52 \times (1.04) = \$0.5408$ or $0.54

Year 3 (two years ago), cost of candy bar = $\$0.5408 \times (1.04)$
$= \$0.562432$ or $0.56

Year 4 (one year ago), cost of candy bar = $\$0.562432 \times (1.04)$
$= \$0.58492928$ or $0.58

Year 5 (present), cost of candy bar = $\$0.58492928 \times (1.04)$
$= \$0.6083264512$ or $0.61

Part B:
The price of the candy bar will continue to increase at a rate of 5% for the next three years.
Therefore, the cost of the candy bar three years from now will be:

$$\$0.6083264512 \times 1.05 \times 1.05 \times 1.05 = \$0.7042139080704 \text{ or } 0.70$$

Year 6 (one year from now), cost of candy bar =
$\$0.6083264512 \times (1 + 0.05) = \$0.63874277376$ *or* $0.64

Year 7 (two years from now), cost of candy bar =
$\$0.6083264512 \times (1 + 0.05) = \$0.670679912448$ *or* $0.67

Year 8 (three years from now), cost of candy bar =
$\$0.6083264512 \times (1 + 0.05) = \$0.7042139080704$ *or* 0.70

Answer #2:

$$\$1.00/1.03/1.03/1.03/1.03 = \$0.888487047 \textit{ or } \$0.89$$

Answer #3:

$$
\begin{aligned}
ADSP &= P \times (i/(1-(1/(1+i)^n))) \\
ADSP &= \$1000 \times (.08/(1-(1/(1+.08)^5))) \\
ADSP &= \$1000 \times (.08/(1-(1/(1.08)^5))) \\
ADSP &= \$1000 \times (.08/(1-(1/1.713824269)) \\
ADSP &= \$1000 \times (.08/(1-.583490395)) \\
ADSP &= \$1000 \times (.08/.416509605) \\
ADSP &= \$1000 \times .192072401 \\
ADSP &= \$192.07
\end{aligned}
$$

Answer #4:

$$
\begin{aligned}
ADSP &= P \times (i/(1-(1/(1+i)^n))) \\
ADSP &= \$500 \times (.04/(1-(1/(1+.04)^3))) \\
ADSP &= \$500 \times (.04/(1-(1/(1.04)^3))) \\
ADSP &= \$500 \times (.04/(1-(1/1.124864000)) \\
ADSP &= \$500 \times (.04/(1-.888996359)) \\
ADSP &= \$500 \times (.04/.111003641) \\
ADSP &= \$500 \times .360348540 \\
ADSP &= \$180.17
\end{aligned}
$$

Answer #5:

Annual Principal Payment = P/n
P = \$175,000
n = 5
Annual Principal Payment = \$175,000/5 = \$35,000

Answer #6:

Annual Principal Payment = P/n
P = \$250,000
n = 10
Annual Principal Payment = \$250,000/10 = \$25,000

Answer #7:

Below is the schedule for the year five maturity at an Interest rate of 5%:

Year	Interest	Principal	Annual Payment
1	$1,500	$0	$1,500
2	$1,500	$0	$1,500
3	$1,500	$0	$1,500
4	$1,500	$0	$1,500
5	$1,500	$30,000	$31,500

Below is the schedule for the year eight maturity at an Interest rate of 7%:

Year	Interest	Principal	Annual Payment
1	$4,900	$0	$4,900
2	$4,900	$0	$4,900
3	$4,900	$0	$4,900
4	$4,900	$0	$4,900
5	$4,900	$0	$4,900
6	$4,900	$0	$4,900
7	$4,900	$70,000	$74,900

Complete the Annual Debt Service Payment schedule for the whole bond by adding together the two above schedules:

Year	Interest	Principal	Annual Payment
1	$6,400	$0	$6,400
2	$6,400	$0	$6,400
3	$6,400	$0	$6,400
4	$6,400	$0	$6,400
5	$6,400	$30,000	$36,400
6	$6,400	$0	$6,400
7	$6,400	$70,000	$76,400

Answer #8:

Loan A:

$$ADSP = \$50,000 \times (.10/(1 - (1/(1+.10)^4)))$$
$$ADSP = \$50,000 \times (.10/(1 - (1/(1.10)^4)))$$
$$ADSP = \$50,000 \times (.10/(1 - (1/1.464100000))$$
$$ADSP = \$50,000 \times (.10/(1 - .683013455))$$
$$ADSP = \$50,000 \times (.10/.316986545)$$
$$ADSP = \$50,000 \times .315470803$$
$$ADSP = \$15,773.54$$

Year	Prior Balance	Interest	–	Total Annual Payment	=	Principal Payment
1	$50,000	$5,000	–	$15,773.54	=	$10,773.54
2	$39,226.46	$3,922.65	–	$15,773.54	=	$11,850.89
3	$27,375.57	$2,737.58	–	$15,773.54	=	$13,035.96
4	$14,339.61	$1,433.96	–	$15,773.54	=	$14,339.61
TOTAL		$13,094.19	–	$63,094.16	=	$50,000

The total amount of money paid by the borrower over the life of the loan = $63,094.16

Loan B:

$$ADSP = \$50,000 \times (.08/(1 - (1/(1 + .08)^6)))$$
$$ADSP = \$50,000 \times (.08/(1 - (1/(1.08)^6)))$$
$$ADSP = \$50,000 \times (.08/(1 - (1/1.586874323))$$
$$ADSP = \$50,000 \times (.08/(1 - .630169627))$$
$$ADSP = \$50,000 \times (.08/.369830373)$$
$$ADSP = \$50,000 \times .216315386$$
$$ADSP = \$10,815.77$$

Year	Prior Balance	Interest	–	Total Annual Payment	=	Principal Payment
1	$50,000	$4,000	–	$10,815.77	=	$6,815.77
2	$43,184.23	$3,454.74	–	$10,815.77	=	$7,361.03
3	$35,823.20	$2,865.86	–	$10,815.77	=	$7,949.91
4	$27,873.29	$2,229.86	–	$10,815.77	=	$8,585.91
5	$19,287.38	$1,542.99	–	$10,815.77	=	$9,272.78
6	$10,014.60	$801.17	–	$10,815.77	=	$10,014.60
TOTAL		$14,894.61	–	$64,894.61	=	$50,000

The total amount of money paid by the borrower over the life of the loan = $64,894.61

Loan A, with a total cost of $63, 094.16, would be less costly than Loan B, with a total cost of $64,894.61.

6 Financial Feasibility

INTRODUCTION

The first five chapters in this series constitute the first five steps in developing the financial information necessary for undertaking projects where income is associated with the delivery of certain utility services, whether the utility is publicly or privately owned. This includes projects in the following utility sectors: water, wastewater, solid waste, and certain types of energy efficiency projects. Those five chapters and this one are written from the perspective of a utility executive who needs a project to improve his system and is interested in learning the financial techniques necessary to obtain the money to undertake the project.

After introducing the concept of project finance (Chapter 1), the first step was the presentation of methods for measuring income (Chapter 2 – Measuring Income). Methods for maximizing income were presented in the succeeding chapter (Chapter 3 – Maximizing Cash Available for Debt Service). It was then emphasized that the maximization of income was of paramount importance because if income exceeded operating expenses, needed projects could be undertaken to improve utility systems by having the utility incur debt. If they had excess income, they would be free of a pernicious reliance on grants. Excess income can be used to pay annual debt service. The next chapter, therefore, was devoted to the explanation of the basic concepts of loans and debt (Chapter 4 – Loan Basics).

Once the basic concepts of borrowing are understood, the next step is to be able to evaluate various types of loans in order to determine which available loan is best for a particular utility. Techniques for evaluating loans, therefore, were presented in the following chapter (Chapter 5 – Project Valuation).

This sixth chapter uses all of the information presented in the preceding five chapters to illustrate which projects are financially feasible for a utility to undertake. It demonstrates both how the largest possible project could be undertaken at the lowest possible cost as well as how a utility's excess operating income can be used to improve the system.

CHAPTER CONTENTS

In this chapter, the relationship between the size of a project and the amount of Cash Available for Debt Service will be discussed. In specific, the following topics will be covered:

- The Relationship between Project Size and Annual Payment
- Lengthening Maturities
- Calculating Project Size using Cash Available for Debt Service
- Coverage Ratios

Introduction

In this module, the relationship between the size of a project and the amount of Cash Available for Debt Service (CADS) will be discussed, specifically:

- Relationship between project size and annual payment
- Lengthening maturities
- Calculating project size using CADS
- Coverage ratios

2

ANNUAL PAYMENTS AND PROJECT SIZE

One of the questions this chapter will answer is:

> "How big a project can my utility afford?"

Annual Payments and Project Size

"How big of a project can my utility afford?"

- The purpose of this question is to learn how much debt a utility system can possibly undertake at a given time
- A quick review of CADS:

CADS = Regular Income – Total Cash Expenses

- In other words, CADS equals the excess cash not needed for the operation and maintenance of the system
- To maximize CADS:
  - Improve bills and collections
  - Adopt consumption based tariffs
  - Replace general subsidies with targeted subsidies
  - Reduce labor, power, and chemical costs
  - Eliminate leaks

3

At first, this may seem odd. Why would one determine the size of a project that was needed by using the amount of Cash Available for Debt Service? It seems, on its surface, as if the one posing the question is willing to build any kind of project – needed or unneeded – and is only interested in knowing how large a project can be undertaken. That is because the purpose of this question is more for instruction and illustration than for real-world, project estimation purposes. The better way of describing the purpose of this question is to learn how to determine how much debt a utility system can possibly undertake at a given time.

In Chapter 2, the method of calculating Cash Available for Debt Service was presented. Cash Available for Debt Service is the difference between recurring cash income and cash expenses. In short, it is the excess cash. It is the cash not needed for the operation and maintenance of the system.

In Chapter 3, strategies for maximizing the Cash Available for Debt Service were discussed. Maximizing Cash Available for Debt Service is done by improving billings and collections, adopting consumption based tariffs, replacing general subsidies with targeted subsidies, and, most importantly, by reducing labor, power, and chemical costs as well as eliminating leaks. When all of these measures have been implemented, the utility should be generating the maximum income which, to the extent it exceeds operating expenses, is Cash Available for Debt Service.

Now, after calculating the maximum amount of CADS a utility system can generate, the amount of debt that can be supported by that amount of cash flow can be calculated as well.

Let us say that, as an example, a utility generates $1,000 per year in Cash Available for Debt Service. If that country had a 0% interest loan program for 100 years, it is easy to see that the utility could borrow $100,000. This is true because the utility can pay back $1,000 per year for 100 years.

Annual Payments and Project Size

- After calculating the maximum amount of CADS a utility system can generate, the amount of debt that can be supported by that amount of cash flow can be calculated as well

For example:

- A utility generates $1,000/yr in CADS
- If the country the utility operates within had a 0% interest loan program for 100 years, the utility could borrow up to $100,000; paid back at $1,000/year

4

Not many countries have 0% interest loan programs, however; and none of them have 100 year loan programs!

If there were a 20 year loan program, the utility could borrow $20,000. If there were a ten year loan program, the utility could borrow $10,000. If there were a five year loan program, the utility could only borrow $5,000. In short, this is why long loan maturities are so important to utility executives. In the example here, if the term of available loans were only increased from five to ten years, the size of a project which a utility system could undertake would double.

This is why it is so important for government officials to work towards lengthening maturities. A water system may need a $20,000 project very badly, but if the maximum maturity available is only five years, then they clearly cannot undertake the project if their Cash Available for Debt Service is only $1,000. This leads us to the next subject.

Annual Payments and Project Size

To further illustrate:

- Assuming a 0% interest rate, the utility could borrow:
  20 year loan ⟹ $20,000
  10 year loan ⟹ $10,000
  5 year loan ⟹ $5,000
- As one can see, if the term of available loans were only increased from 5 to 10 years, the project size a utility system could undertake would double
- A water system may need a $20,000 project, but if the maximum maturity available is only 5 years, then they clearly cannot undertake the project with a CADS of $1,000
- The previous point further demonstrates the importance for government officials to work towards lengthening maturities

5

LENGTHENING MATURITIES

One hundred years ago in the developed countries it was easy for banks to make long-term (20–30 year) loans. That is because people who deposited money into the banks had very little else to do with it. There were relatively very few investment opportunities 100 years ago, as opposed to today where there are a myriad of such opportunities. As a result of the fact that there were few investment opportunities, people left their funds in banks for many years. The banks, in turn, were able to make long-term loans because they had plenty of long-term deposits with which to make them. Now, because there are so many investment opportunities, people no longer leave their funds in banks for prolonged periods of time. As a result, commercial banks do not have funds with which to make long-term fixed-rate loans.

Lengthening Maturities

In today's financial world, there are countless investment opportunities and, as a result, people no longer leave their funds in banks for a prolonged period of time leading to a reduction in the amount of funds commercial banks have available to make long-term fixed-rate loans.

6

In developing countries, which may not have many alternative investments, the commercial banks have other problems that inhibit their ability to make long-term loans. The first of these is their inability to measure long-term financial risk.

Lengthening Maturities

In developing countries, where investment alternatives may be limited, commercial banks have other problems that inhibit their ability to make long term loans, such as:

- Inability to measure long-term financial risk
  - Few businesses keep financial books and records in accord with internationally accepted accounting standards
  - The few records kept and shared with lenders are often entirely obscure

7

In developing countries, few business enterprises keep financial books and records in accord with internationally accepted accounting standards. What few records businesses do keep, and are willing to share with lenders, may be wholly opaque. Lenders have come to expect few meaningful financial records and are suspicious of the ones they see. Thus, they have no basis upon which to make a long-term loan decision.

Another problem is fear of a liquidity crisis. In less developed countries, local economies may be fragile. Rumors about certain government actions, or rumors about a bank itself, may send depositors streaming in to collect their deposited funds. Funds that are tied up in long-term commitments cannot be paid out to depositors. This (i.e. not having immediately available funds) usually worsens whatever crisis is afoot.

A third problem that inhibits banks' long-term lending is inflation and fear of inflation. Economies that are moving along with 5%, 10%, or even 20% inflation rates can suddenly implode, and their inflation rates burgeon to 100%+ or even 1000%+ per year. If a bank loaned a company $1 million for ten years and the average inflation rate was 200% over that ten year period, the bank would be bankrupt.

Lengthening Maturities

Problems commercial banks may experience in developing countries (cont.):

- Fear of a liquidity crisis
  - Fragile local economies
  - Negative rumors about government and banks
- Inflation and fear of inflation
  - Banks will need to issue variable interest rates tied to inflation

8

Now, it is possible for banks to make loans with variable interest rates with the rates tied to inflation. This would mean that a loan with a 10% rate today might automatically have a 50% rate next month. As a result, the bank will be getting more dollars, or pesos, or rubles so that they can meet their own obligations (such as the rates on deposits) that are tied to inflation.

Now, in developed countries, the only financial institutions with appetites for long-term debt are life insurance companies and pension funds. Pension funds need long-term investments because the money they are taking in today from young workers will not be paid out to them for 20–40 years in the form of pension payments. Hence, pension funds need investments that have commensurately long-terms.

Life insurance companies are in a similar situation. These companies receive premium income today from policyholders who will not die for 20 – 40+ years. Life insurance companies have specialized accountants – called actuaries – who estimate the lifespan of their policyholders. The actuaries estimate the length of time between the receipt of payment of premiums and the payment of a claim. Since these periods can run to 40+ years, the companies must look for long-term investments to make so that they can earn interest on the funds.

Pension funds and life insurance companies are, therefore, the key institutions in developing the demand side of the market for long-term debt which environmental infrastructure projects need.

Lengthening Maturities

In developed countries, life insurance companies and pension funds represent the few financial institutions that posses an interest in long-term debt. These key institutions will develop the demand side of the market for long-term debt which environmental infrastructure projects need.

9

CALCULATING PROJECT SIZE

Once Cash Available for Debt Service is known, calculating project size can be relatively simple.

To begin, we can assume that Cash Available for Debt Service equals $150,000. Three questions must then be answered:

1) What loan terms are available?
2) At what rates of interest?
3) Are payments by the level payment method or the level principal method?[1]

Calculating Project Size

Once CADS is known, three questions must be answered:

1. What loan terms are available?
2. At what rates of interest?
3. Are payments by the level payment method or the level principal payment method*

*Refer to Module V – Project Valuation

10

1 See Chapter 5 – Project Valuation.

Let us say that loans with five, seven, and ten year terms are available. Let us say that the respective interest rates on these loans are 6%, 7%, and 9%.

For every $1,000 borrowed, using the level principal payment method, the first annual payment on each of the respective loans would be:

Five years/6%	–	$260
Seven years/7%	–	$213
Ten years/9%	–	$190

Calculating Project Size

For example, let us assume:

- CADS = $150,000
- Loans with 5,7, and 10 year terms are available
- The respective interest rates are 6%, 7%, and 9%
- $1,000 will be borrowed at a time using the **Level Principal Payment** method

The first annual payment on each of the respective loans would be:

5 yr loan at 6% ➡ $260
7 yr loan at 7% ➡ $213
10 yr loan at 9% ➡ $190

11

If Cash Available for Debt Service is $150,000, then the maximum debt which this utility could assume – with the five-year/6% loan – would be $596,923. (150,000/260 * 1000) With a seven-year/7%, loan the maximum project size would be $704,225. (150,000/213 * 1000) With the ten-year/9% loan, the maximum project size would be $789,474. (150,000/190 * 1000)

Calculating Project Size

Example (cont.)

Using the **Level Principal Payment** Method

If CADS is $150,000, the maximum debt this utility could assume is as follows:

5 yr loan at 6% ➡ ($150,000/$260 * $1,000) ➡ $596,923

7 yr loan at 7% ➡ ($150,000/$213 * $1,000) ➡ $704,225

10 yr loan at 9% ➡ ($150,000/$190 * $1,000) ➡ $789,474

12

Now, let us make the same calculations using the level payment method.

For every $1,000 borrowed, using the level payment method, the annual payments on each of the respective loans would be:

5 years/6%	–	$237
7 years/7%	–	$186
10 years/9%	–	$156

Calculating Project Size

Example (cont.)

Using the **Level Payment** Method

For every $1000 borrowed, the annual payment method on each of the respective loans would be:

5 yr loan at 6% ➡ $237

7 yr loan at 7% ➡ $186

10 yr loan at 9% ➡ $156

13

If Cash Available for Debt Service is $150,000, then the maximum debt which this utility could assume – with the five-year/6% loan – would be $632,911. (150,000/237 * 1000) With a seven-year/7%, loan the maximum project size would be $806,452. (150,000/186 * 1000) With the ten-year/9% loan, the maximum project size would be $961,538. (150,000/156 * 1000)

Calculating Project Size

Example (cont.)

Using the **Level Payment** Method

If CADS is $150,000, the maximum debt this utility could assume is as follows:

5 yr loan at 6% ➡ ($150,000/$237 * $1,000) ➡ $632,911

7 yr loan at 7% ➡ ($150,000/$186 * $1,000) ➡ $806,452

10 yr loan at 9% ➡ ($150,000/$156 * $1,000) ➡ $961,538

14

As is evident, if loans are available with longer terms, even with moderately higher interest rates, larger projects can be undertaken. Likewise, if loans using the level payment method are available, even larger projects can be undertaken.

Calculating Project Size

As one can see, loans with longer terms, even with moderately higher interest rates, allow for larger projects to be undertaken. Likewise, loans using the level payment method will yield even larger projects.

15

Now, let us say that the minimum size of a project that a utility must undertake is $1,000,000. In this case, if the utility only has $150,000 of Cash Available for Debt Service, and if only the above loans are available, then the utility must seek a grant for the balance of the project cost. In this case, the amount of the grant is the difference between the required project cost ($1,000,000) and the maximum project size that can be undertaken as debt.

Calculating Project Size

Let us assume that the minimum size of a project that a utility must undertake is $1,000,000. Building on the previous example, the utility only has $150,000 of CADS, and with the previous loans available (5yr/6%, 7yr/7%, and 10yr/9%), the utility must seek a grant to cover the additional project cost.

In this case, the amount of the grant is the difference between the required project cost ($1,000,000) and the maximum project size that can be undertaken as debt.

16

Using the level principal payment loans, the grants would have to be:

Five years/6%	–	$403,077
Seven years/7%	–	$295,775
Ten years/9%	–	$210,526

Calculating Project Size

Using the **Level Principle Payment** Method, the grants would have to be:

5 yr loan option at 6% ➡ $403,077
7 yr loan option at 7% ➡ $295,775
10 yr loan option at 9% ➡ $210,526

17

Using the level payment method, the amount of the respective grants would have to be:

Five years/6%	–	$367,089
Seven years/7%	–	$193,548
Ten years/9%	–	$38,462

The obvious lesson here is that it is far easier to get a grant for $38,462, representing 3.8% of total project cost, than it would be to get a grant of $403,077, representing over 40.3% of total project cost.

So, the lesson is: Maximize Cash Available for Debt Service.

COVERAGE RATIOS

There is one small problem that can affect the calculations above. That is, lenders often require that borrowers have a cash reserve, which acts as a cushion, in case next year's Cash Available for Debt Service falls short of this year's Cash Available for Debt Service. This cushion is called "coverage". It is usually expressed as a ratio. Hence, it is most often referred to as a required "coverage ratio".

Coverage Ratios

Lenders often require borrowers to have a cash reserve, serving as a cushion in the case that next year's CADS falls short of the present year's CADS. This cushion is usually expressed as a ratio, referred to as a *coverage ratio*.

19

Let us say that a lender requires that a utility have a 105% coverage ratio. That means that their Cash Available for Debt Service must be 5% higher than the annual debt service payments. So, in our example, although the utility has $150,000 of Cash Available for Debt Service, only 95%, or $142,500, can be used to calculate the size of the loan.

Coverage Ratios

Let us assume that a lender requires a utility to have a 105% coverage ratio. This means that their CADS must be 5% higher than the annual debt service payments.

For example: If the utility has a CADS of $150,000, only 95%, or $142,500, can be used to calculate the size of the loan.

20

In this case, we must perform the above calculations using $142,500 as Cash Available for Debt Service.

Using the level principal payment method, the respective maximum project sizes would be:

Five years/6%	–	$548,077
Seven years/7%	–	$669,014
Ten years/9%	–	$750,000

Coverage Ratios

Example (cont.)

Using the **Level Principal Payment** Method

If there is a 105% coverage ratio, CADS is now $142,500, so the maximum debt this utility could assume is as follows:

5 yr loan at 6% ➡ ($142,500/$260 * $1,000) ➡ $548,077

7 yr loan at 7% ➡ ($142,500/$213 * $1,000) ➡ $669,014

10 yr loan at 9% ➡ ($142,500/$190 * $1,000) ➡ $750,000

21

Using the level payment method, the respective maximum project sizes would be:

Five years/6%	–	$601,266
Seven years/7%	–	$766,129
Ten years/9%	–	$913,462

Coverage Ratios

Example (cont.)

Using the **Level Payment** Method

If there is a 105% coverage ratio, CADS is now $142,500, so the maximum debt this utility could assume is as follows:

5 yr loan at 6% ➡ ($142,500/$237 * $1,000) ➡ $601,266

7 yr loan at 7% ➡ ($142,500/$186 * $1,000) ➡ $766,129

10 yr loan at 9% ➡ ($142,500/$156 * $1,000) ➡ $913,462

22

Now, we can recalculate the amount of grants necessary with the coverage ratio requirement with loans using the level principal payment method, as follows:

Five years/6%	–	$451,923
Seven years/7%	–	$330,986
Ten years/9%	–	$250,000

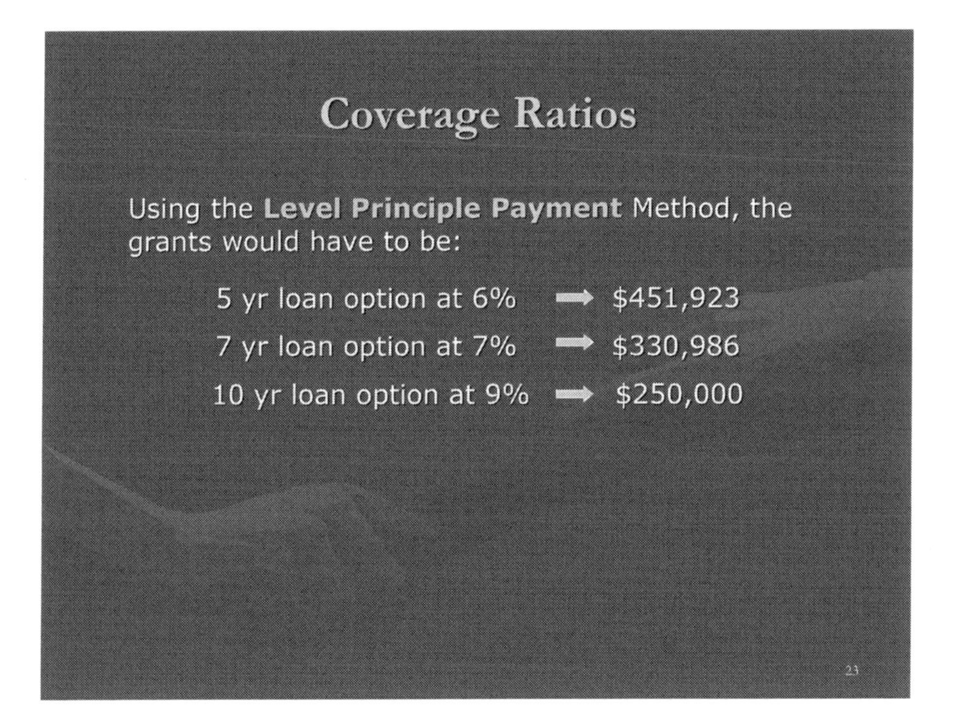

We can also recalculate the amount of grants necessary with the coverage ratio requirement with loans using the level payment method, as follows:

Five years/6%	–	$398,734
Seven years/7%	–	$233,871
Ten years/9%	–	$86,538

Again, it is obvious that a utility is more likely to get a grant of $86,538, representing 8.7% of total project cost, than it is to get a grant of $451,923, representing 45.2% of project cost.

Coverage Ratios

Using the **Level Payment** Method, the grants would have to be:

5 yr loan option at 6% ➡ $398,734
7 yr loan option at 7% ➡ $233,871
10 yr loan option at 9% ➡ $86,538

Once again, there is a greater likelihood that a utility would receive a grant of $86,538 (8.7% of the total project cost), as opposed to $451,923 (45.2%).

24

So, the lessons of this chapter, and of the preceding five chapters, are:

1. Minimize grant requests by maximizing the amount of a project that can be undertaken with debt.
2. Maximize Cash Available for Debt Service.
3. Take the loan with the longest possible term.
4. Use the level payment method wherever possible.

Conclusion

The lessons of this module, and of the preceding five modules are:

- Minimize grant requests by maximizing the amount of a project that can be undertaken with debt
- Maximize CADS
- Select the loan with the longest possible term
- Use the level payment method wherever possible

25

Questions and Answers

Question #1:

A utility generates $200,000 per year in CADS. The bank has three different loan options with three, six, and nine year terms. The respective interest rates are 4%, 5%, and 8%.

Part A: For every $1,000 borrowed, using the level principal payment method, calculate the first annual payment on each of the respective loans.

Part B: After determining the first annual payment, using the level principal payment method, calculate the maximum debt, or maximum project size, that the utility can undertake for each loan alternative.

Part C: The minimum project size that this utility must undertake is $1.5 million. In this case, calculate the amount of grants the utility must receive to cover the remaining balance of the project cost.

Part D: The lender now requires the utility to have a 110% coverage ratio. Continuing to use the level principal payment method, recalculate the new maximum debt, or maximum project size, that the utility can undertake for each loan alternative.

Part E: After determining the new maximum project size, factoring in the 110% coverage ratio and using the level principal payment method, recalculate the amount of grants the utility must receive to cover the remaining balance of the project cost.

Question #2:

A utility generates $75,000 per year in CADS. The bank has three different loan options with eight, ten, and 11 year terms. The respective interest rates are 7%, 9%, and 12%.

Part A: For every $1,000 borrowed, using the level payment method, calculate the first annual payment on each of the respective loans.

Part B: After determining the first annual payment, using the level payment method, calculate the maximum debt, or maximum project size, that the utility can undertake for each loan alternative.

Part C: The minimum project size that this utility must undertake is $500,000. In this case, calculate the amount of grants the utility must receive to cover the remaining balance of the project cost.

Part D: The lender now requires the utility to have a 103% coverage ratio. Continuing to use the level payment method, recalculate the new maximum debt, or maximum project size, that the utility can undertake for each loan alternative.

Part E: After determining the new maximum project size, factoring in the 103% coverage ratio and using the level payment method, recalculate the amount of grants the utility must receive to cover the remaining balance of the project cost.

Answer #1:

Part A:
For every $1,000 borrowed, the first annual payment for each loan option is as follows:

Three year loan at 4% interest – $373.33

Six year loan at 5% interest – $216.67

Nine year loan at 8% interest – $191.11

Part B:
The maximum debt, or maximum project size, that the utility can expect from each loan option is as follows:

CADS = $200,000

Three year loan at 4% interest – (200,000/373.33 * 1,000) = $535,719.07

Six year loan at 5% interest – (200,000/216.67 * 1,000) = $923,062.72

Nine year loan at 8% interest – (200,000/191.11 * 1,000) = $1,046,517.71

Part C:
The amount of grants needed to cover the remaining balance of the project cost is as follows:

Three year loan at 4% interest – $1,500,000 − $535,719.07 = $964,280.93

Six year loan at 5% interest – $1,500,000 − $932,062.72 = $567,937.28

Nine year loan at 8% interest – $1,500,000 − $1,046,517.71 = $453,482.29

Part D:
With a 110% coverage ratio, the new maximum debt, or maximum project size, that the utility can expect from each loan option is as follows:

Three year loan at 4% interest – (180,000/373.33 * 1,000) = $482,147.16

Six year loan at 5% interest – (180,000/216.67 * 1,000) = $830,756.45

Nine year loan at 8% interest – (180,000/191.11 * 1,000) = $941,865.94

Part E:
Taking into account the 110% coverage ratio, the amount of grants needed to cover the remaining balance of the project cost is as follows:

Three year loan at 4% interest – $1,500,000 − $482,147.16 = $1,017,852.84

Six year loan at 5% interest – $1,500,000 − $830,756.45 = $669,243.55

Nine year loan at 8% interest – $1,500,000 − $941,865.94 = $558,134.06

Answer #2:

Part A:
For every $1,000 borrowed, the first annual payment for each loan option is as follows:

Eight year loan at 7% interest – $167.47

Teb year loan at 9% interest – $155.82

11 year loan at 12% interest – $168.42

Part B:
The maximum debt, or maximum project size, that the utility can expect from each loan option is as follows:

CADS = $75,000

Eight year loan at 7% interest – (75,000/167.47 * 1,000) = $447,841.40

Ten year loan at 9% interest – (75,000/155.82 * 1,000) = $481,324.61

11 year loan at 12% interest – (75,000/168.42 * 1,000) = $445,315.28

Part C:
The amount of grants needed to cover the remaining balance of the project cost is as follows:

Eight year loan at 7% interest – $500,000 − $447,841.40 = $52,158.60

Ten year loan at 9% interest – $500,000 − $481,324.61 = $18,675.39

11 year loan at 12% interest – $500,000 − $445,315.28 = $54,684.72

Part D:
With a 103% coverage ratio, the new maximum debt, or maximum project size, that the utility can expect from each loan option is as follows:

Eight year loan at 7% interest – (72,750/167.47 * 1,000) = $434,406.16

Ten year loan at 9% interest – (72,750/155.82 * 1,000) = $466,884.87

11 year loan at 12% interest – (72,750/168.42 * 1,000) = $431,955.82

Part E:
Taking into account the 103% coverage ratio, the amount of grants needed to cover the remaining balance of the project cost is as follows:

Eight year loan at 7% interest – $500,000 − $434,406.16 = $65,593.84

Ten year loan at 9% interest – $500,000 − $466,884.87 = $33,115.13

11 year loan at 12% interest – $500,000 − $431,955.82 = $68,044.18

7 Alternative Finance Options

INTRODUCTION

The preceding six chapters constitute the first six steps in developing the financial information necessary for undertaking projects where income is associated with the delivery of certain utility services, whether the utility publicly or privately owned. This includes projects in the following utility sectors: water, wastewater, solid waste, and certain types of energy efficiency projects. Those six chapters are written from the perspective of a utility executive who needs a project to improve his system and is interested in learning the financial techniques necessary to obtain the money to carry out the project.

After introducing the concept of project finance (Chapter 1), the first step was the presentation of methods for measuring income (Chapter 2 – Measuring Income). Methods for maximizing income were presented in the succeeding chapter (Chapter 3 – Maximizing Cash Available for Debt Service). It was then emphasized that the maximization of income was of paramount importance because if income exceeded operating expenses, needed projects could be undertaken to improve utility systems by having the utility incur debt. If they had excess income, they would be free of a pernicious reliance on grants. Excess income can be used to pay annual debt service. The next chapter, therefore, was devoted to the explanation of the basic concepts of loans and debt (Chapter 4 – Loan Basics).

Once the basic concepts of borrowing are understood, the next step is to be able to evaluate various types of loans in order to determine which available loan is best for a particular utility. Techniques for evaluating loans, therefore, were presented in the following chapter (Chapter 5 – Project Valuation).

The next chapter used all of the information presented in the preceding five chapters to illustrate which projects were financially feasible for a utility to undertake. It demonstrated both how the largest possible project could be undertaken at the lowest possible cost and how a utility's excess operating income could be used to improve the system (Chapter 6 – Financial Feasibility).

Now, in this chapter, the perspective shifts. Financial concepts will no longer be presented from the utility executive's perspective. Rather, in this chapter, financial concepts will be presented from the perspective of a government official whose responsibility it is to manage or administer a program to provide funds for municipal utility services. In other words, in the preceding six chapters, project finance has been discussed from the perspective of one needing funds (for a particular project). In this chapter, the concepts of project finance will be discussed from the perspective of a government official who has funds, or whose job it is to seek funds for utility projects in his country. It is this official's responsibility to provide the best quality of utility services to the largest amount of people, by providing funds to utility operators for projects, which improve that utility's services to its users.

CHAPTER CONTENTS

This chapter contains information on the four ways to finance projects: grants, soft or subsidized loans, market rate loans, and loan guaranties. There is a fifth way to finance projects, but it is unique to the private sector. That way is called "equity", or equity financing. Some people confuse equity and grants. These concepts should not be confused. Grants do not need to be repaid. Equity, on the other hand, requires the greatest payback. In fact, that is precisely the reason why equity is not considered in this chapter. The payback on equity is so high that it is virtually impossible to finance utility projects by this method.

It needs to be emphasized that there are, in reality, only three ways to finance any type of project. They are: grants, loans, and equity. As stated above, equity need not be discussed here because utility projects are virtually impossible to fund with equity. In addition, subsidized loans are, in reality, just a mix of grants and loans. Finally, market rate loans and loan guaranties are just variations – albeit very important ones – of the same basic concept.

It also needs to be re-emphasized that there are, in reality, only three ways to finance any type of project. This is because the term "innovative finance" is often used in international government circles as if it were some mysterious "other" way to finance. It is not. All finance is grants, loans, or equity. When officials use the term "innovative finance", it usually means that they are trying to identify a donor to give them a grant.

So, this chapter will present three matters. It will present the concepts of grants, soft or subsidized loans, market rate loans, and loan guaranties. Second, it will present the pros and cons of each of these methods of financing. And, third, it will compare these four finance techniques from the perspective of how many projects can be done by each method with identical sums of money.

FOUR TECHNIQUES FOR FINANCING PROJECTS

There are four ways to finance projects:

1) *Grants*: sums of money given or awarded to finance a particular activity or project, which do not need to be repaid.
2) *Subsidized Loans* (below market interest rates): a loan made to a qualified borrower at below the current market rate of interest.
3) *Market Rate Loans*: a loan made to a qualified borrower at the current market interest rate.
4) *Loan Guaranties*: a promise from a guarantor to make payment to the lenders in the case of nonpayment by the borrowers.

Types of Financing

There are five ways to finance projects:

- Grants
- Bonds
- Subsidized (sub-market rate) Loans
- Market Rate Loans
- Loan Guaranties

2

In creating or establishing a financial system for the funding of environmental infrastructure projects, governments must choose which of the above, or which combination of the above, funding mechanisms it will use to finance projects. Each financing option has positive and negative aspects that must be considered when deciding which to incorporate into an environmental finance program. The most important factor for government officials to consider when making project financing decisions is the amount of projects that can be funded with the limited resources available; this will be referred to as *efficiency.*

Efficiency: with respect to project finance, is the number of projects that can be funded with fixed or limited capital resources; high efficiency occurs when the most projects are funded for the least amount of capital.

The following example will help to demonstrate the impact on efficiency under different finance option scenarios:

Assume the existence of a "National Environmental Fund" (NEF, which is a government agency that has seed money in the amount of $100,000,000. The effect of the NEF's project finance decisions will be simulated over a ten year period, starting in Year 0. Each project is assumed to cost $5,000,000.

GRANTS

Grants

- Grants are sums of money awarded to finance a particular activity or facility
- Grant awards do not need to be paid back

3

Since grants are not repaid, they improve cash flow (cash available) for the environmental project; money that would have been spent for loan repayments can be put to other uses for the project. Grants are made to facilitate projects that are otherwise not affordable. These benefits come at some costs to the project managers since grantors often attach specific project requirements and conditions along with the funding, which limit project managers in how the grant money is used. Projects are often required to meet the specific goals of the granting governmental agency or private organization. From the perspective of the funding agency, a grant is the least efficient use of fixed capital assets since once the grant is made the money is gone and is not repaid.

The NEF managers choose to fund all environmental projects with grants. So, they make 20 grants in Year One. That is all. There are no funds for grants in subsequent years. The following graph shows the impact of making 20 grants in Year One.

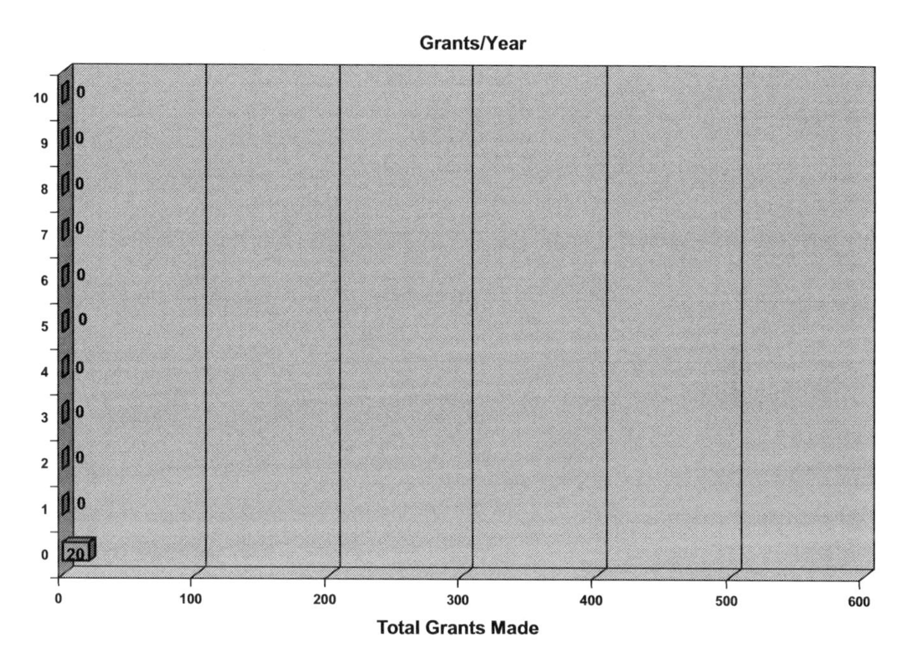

Notice that at the end of Year One, there is no money left in the NEF account to finance future projects. A total of 20 projects were funded with the $100,000,000 in the Fund.

Grants

Pros:

- Do not need to be repaid
- Improve project cash flow by eliminating loan repayments
- Facilitate projects that are otherwise not affordable

4

Grants

Cons:

- Foster dependence
- Encourage overbuilding of projects
- Often disregard costs of operation and maintenance
- Generally, very specific requirements and conditions attached
- Requirement to meet specific goals of the particular federal agency or private organization
- For programs with fixed capital, once grant is made, funds are gone

5

SUBSIDIZED LOANS

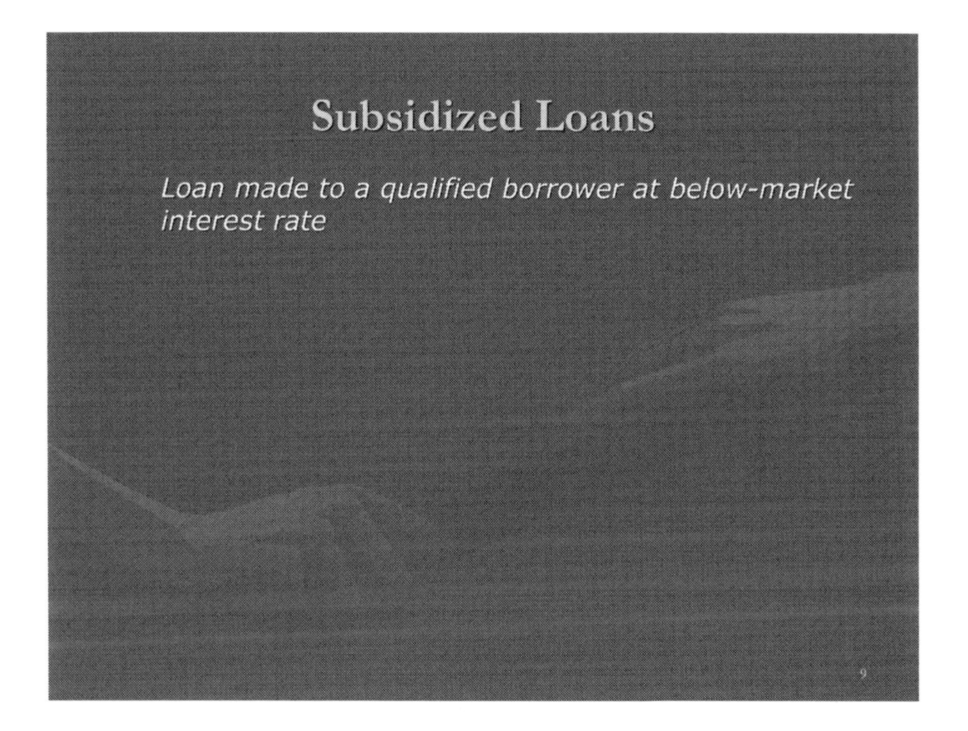

Subsidized loans are loans that have an interest rate that is lower than the market rate of interest. There is a cost to subsidizing loans.

The cost of a subsidized loan is the present value of the difference between the payments on a market rate loan of the same tenor and the subsidized loan. Thus if the payments on a market rate loan would be $10,000 per year for five years and the payments on a subsidized loan are $6,000 for five year, then the cost of the subsidy is the present value of the difference in the two payments ($10,000 – $6,000 = $4,000) each year. The amount of the subsidy can be considered a grant.

(The amount of a subsidy can be considered a grant because if the funds had been loaned out at market rates, the repayments to the NEF would be significantly higher. Thus, the lending institution [here the NEF] "loses" money. Another way of looking at this is in terms of the growth of the NEF's account. If it lends at a low rate, it receives lower repayments; which means that the NEF account grows much slower than it would have normally (i.e. if it had made market rate loans). The shortfall in this growth is the value of the subsidy, which should be considered a grant.)

With subsidized loans, the subsidy reduces the interest rate that borrowers pay on their loans. Lower interest rates result in lower periodic payments, leaving more cash available for the utility, or otherwise enabling it to undertake a larger project than it could have without the lower, subsidized interest rate. Like grants, subsidized loans are used to facilitate projects that would not be eligible for market rate loans due to limited cash availability. To receive a subsidized loan, borrowers must often meet the requirements and conditions set by the government agency or private donor that is providing the subsidy. These loans have a higher efficiency than grants because part of the project's funding is paid back; however, there is still a capital loss to the NEF equal to the amount of the subsidy. As indicated above, the subsidy amount is calculated by subtracting the subsidized interest payments from current market interest rate payments, then multiplying this difference by the outstanding balance of the loan each year, and then discounting each amount back to a present value.

$$(\text{Market Interest Rate} - \text{Subsidized Interest Rate}) * \text{Amount of Loan} = \text{Subsidy}$$

The NEF managers decide to use the $100,000,000 to make subsidized loans to finance all environmental projects. Loans are for a period of five years, repaid according to the ***level principal method****. The interest earned on subsidized loans is assumed to be zero. All money that is received by the NEF as repayment of principal is re-loaned in order to finance more projects. Since all projects require $5,000,000 there may be some money (less than $5,000,000) rolled over to the next year. The simulation occurs over a ten-year period. New loans are made each year from the annual principal repayments. The graph below shows the how many subsidized loans can be made with the same $100,000,000 over a - ten year period.*

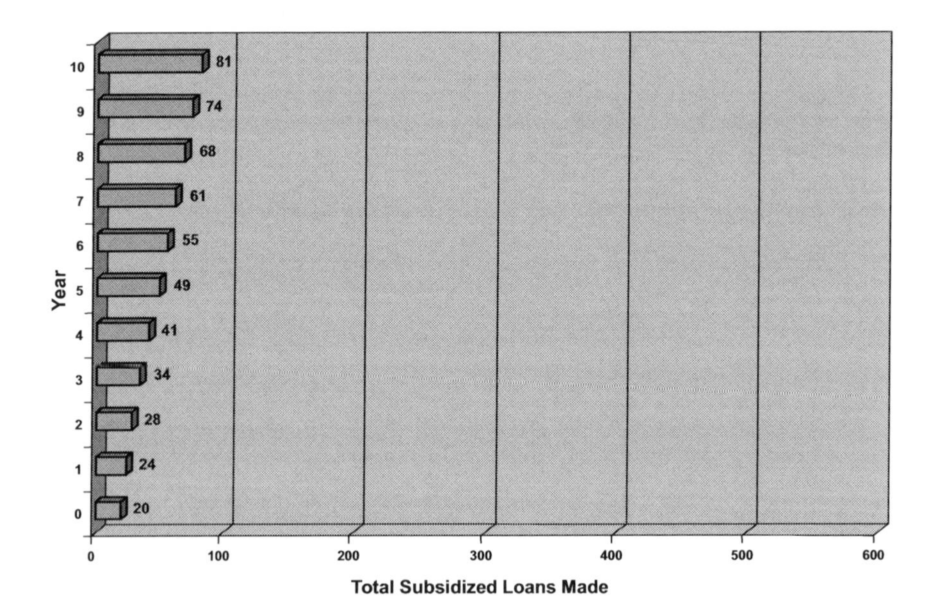

From this table, notice how the same amount of seed capital ($100,000,000) provides financing for 81 projects when using subsidized loans as compared with only 20 projects financed by grants. Here, subsidized loans are funding 405% more projects than grants. So, subsidized loans are much more efficient than grants.

Subsidized Loans

Pros:

- Improve project cash flow by decreasing loan payments

Cons:

- Generally, requirements and conditions attached
- Requirement to meet specific goals of the particular federal agency or private organization
- Encourage financial dependence
- For programs with fixed capital, fewer loans can be made

10

MARKET RATE LOANS

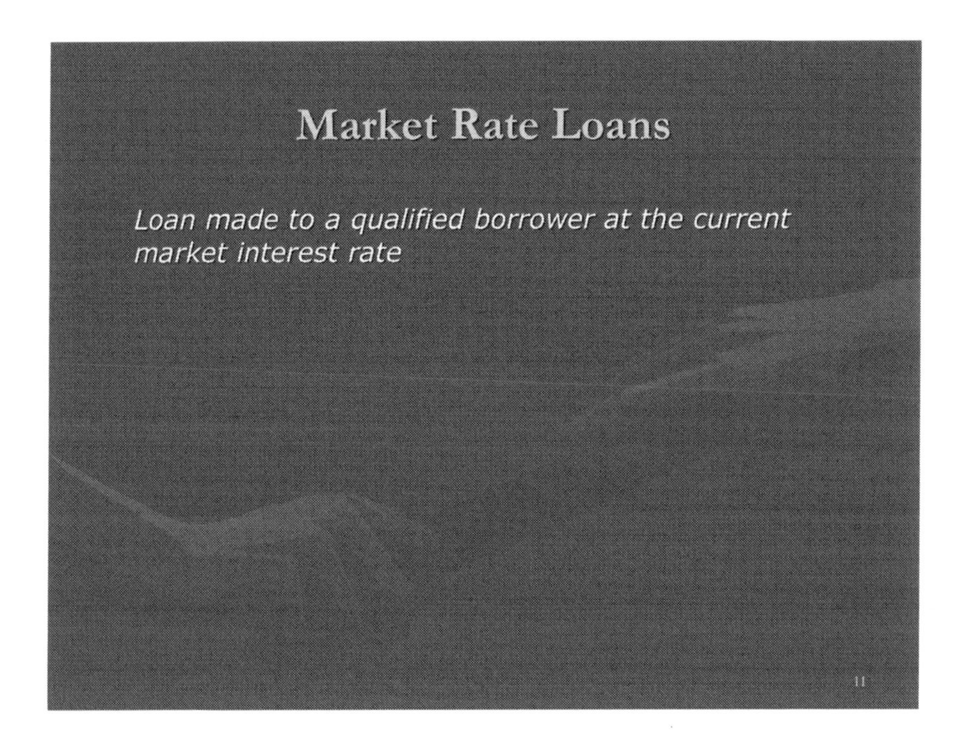

An option for providing funding for more environmental projects than could be achieved with grants or subsidized loans is to provide loans at the market interest rate. Since market rate loans do not provide a subsidy, they usually do not have conditions regarding the way that projects are carried out; in this sense, they are easier for borrowers to obtain than grants and subsidized loans. The higher interest rate (as compared with the subsidized loan) increases the total amount of the loan repayment, which results in higher periodic loan payments. These higher loan payments decrease the amount of available cash flow for the project.

NEF managers now choose to provide all environmental projects financing with market rate loans. The current market rate is 10% for loans of $5,000,000 over a five year period. Payments are calculated based on the level principal method (as in the previous example). The bar graph below illustrates how many projects can be funded with market rate loans.

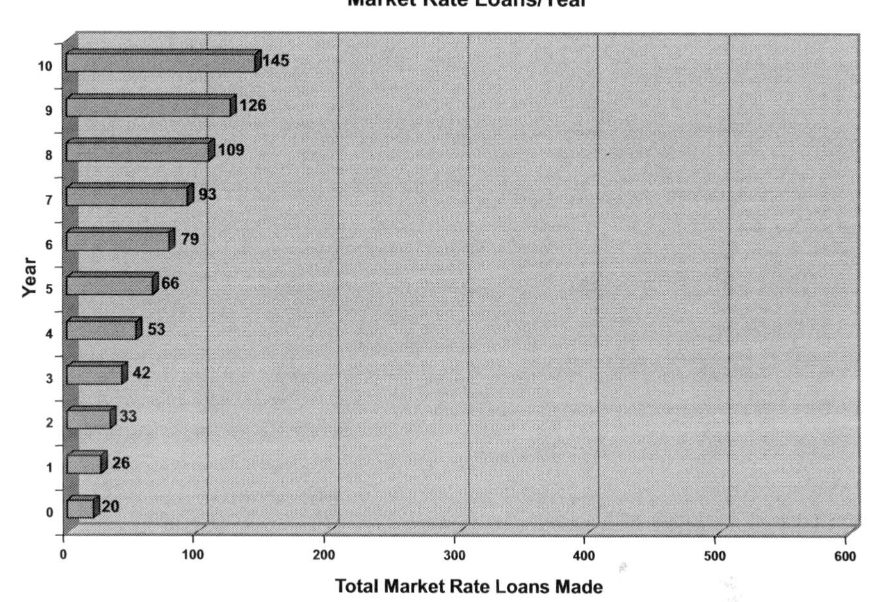

Now that the NEF is no longer providing a subsidy to all of the environmental projects that it is financing, many more projects are eligible for financing over the ten-year period. Both interest earned and principal repayments are re-loaned in order to provide financing for more projects, resulting in 145 projects being funded over the ten-year period. Market rate loans provide a 725% increase in the number of projects financed over grants and an 80% increase in the number of projects financed over subsidized loans.

Market Rate Loans

Pros:

- Easier to obtain than grant, subsidized loan, or loan guarantee
- Generally, has no or limited conditions regarding the way projects are carried out
- For programs with fixed capital, more loans can be made

12

Market Rate Loans

Cons:

- Decrease project cash flow by increasing loan payments
- Requires higher loan repayment

13

MUNICIPAL BONDS

Before we leave Market Rate Loans, we need to say a few words about Municipal Bonds.

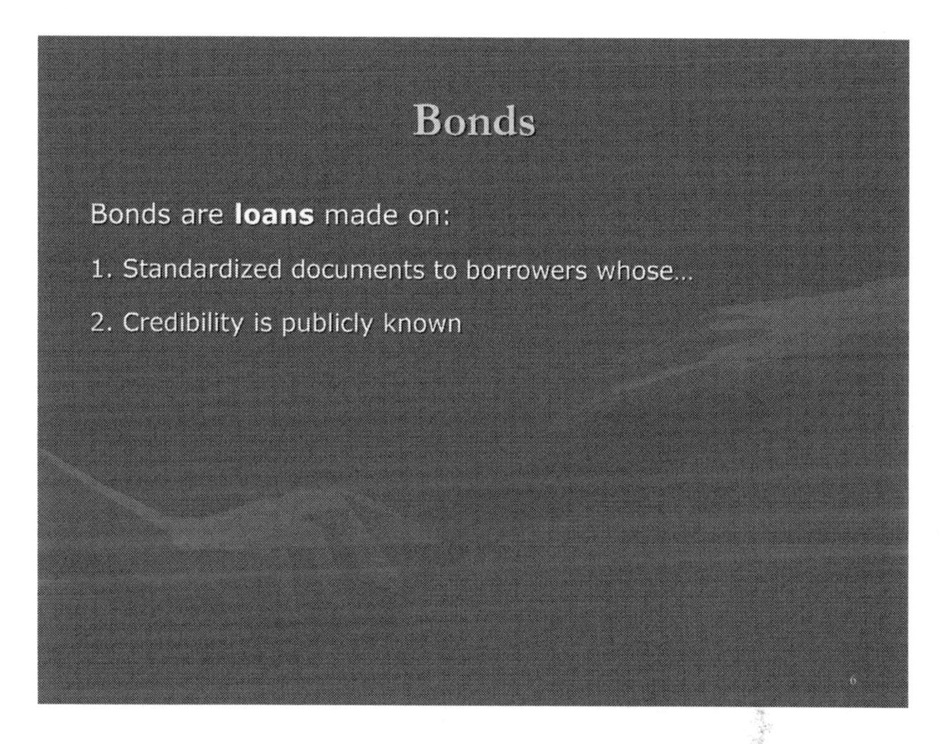

As you can see, Bonds are just a special type of Market Rate Loan. In some countries there are well-developed bond markets. In the United States, the municipal bond market is over $3.8 trillion with some $400 billion of new bonds issued each year. In the US, however, municipal bonds are a special case. For most environmental infrastructure projects, and other public projects, such as schools, hospitals, roads, etc., the interest income on the bonds is exempt from federal taxation, and, in most states, from state income taxation, as well. These tax-exempt bonds comprise 85%–90% of the entire municipal bond market in the US.

Bond markets create great value for borrowers. With thousands of buyers competing for bonds, the lowest possible rates are available. The same occurs with term. As you know, the longer the term of a bond or loan, the lower the annual payment. So, again with thousands of buyers competing for bonds, the longest possible terms are available too.

Bonds

Pros:

- Are very liquid
- Encourage growth of financial markets, especially secondary markets
- Bonds allow borrowing for longer maturities and larger amounts than market rate loans

7

Bonds

Cons:

- Borrowers must submit to public credit analysis
- Borrowers must use internationally accepted accounting standards

8

LOAN GUARANTIES

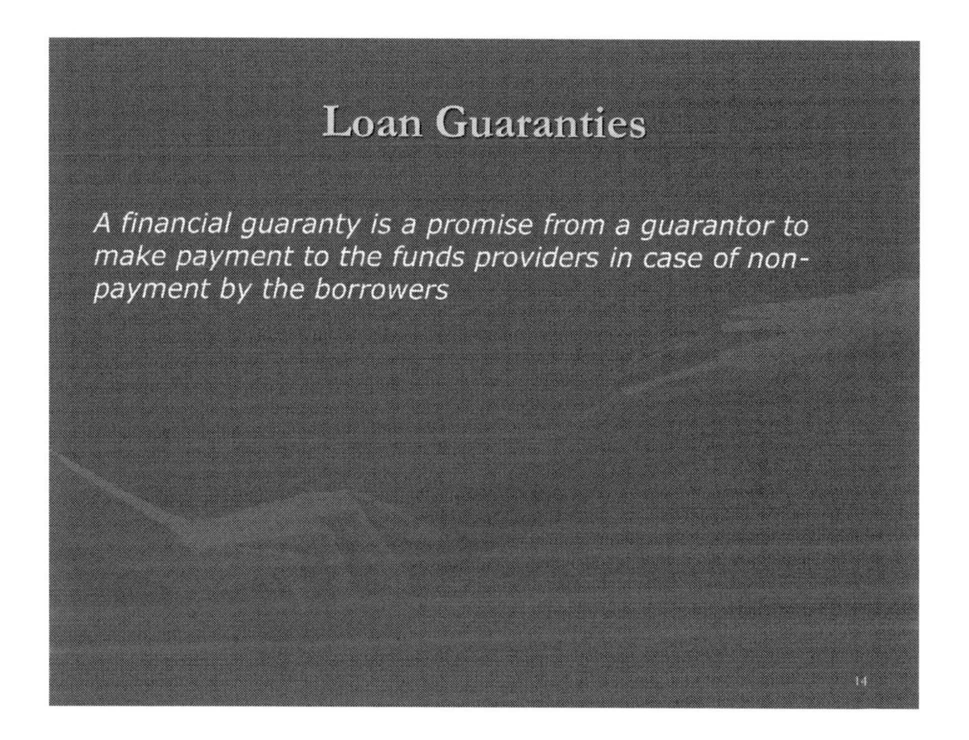

Loan guaranties use the capital assets held in the national Fund as a means of guarantying the repayment of loans made by private banks (or other lenders) to borrowers for environmental projects. The NEF only needs to release money to the lenders in the case of nonpayment by the borrowers. This structure allows the borrowers to negotiate favorable lending terms with the bank, usually resulting in: below market interest rates, longer repayment periods, and larger loan amounts. Borrowers are able to secure lower interest rates, because the guaranty provided by the Fund reduces the credit risk that is borne by the lending bank. Loan guaranties provide more benefits to borrowers as compared with borrower at market rates, although the benefits are not as high as in the case of grants or subsidized loans. Generally, borrowers must comply with the conditions and regulations that are established in the guaranty agreement for the environmental project. These terms and conditions are usually the same for both market rate loans and loan guaranties.

The NEF managers have decided to allow existing banks (or other money lending institutions) to provide the loan funds for the environmental projects. To ensure that the environmental projects are eligible for private financing and receive favorable treatment, the NEF will agree to guaranty the loans. The bar graph below illustrates how many projects can be funded with the same $100,000,000 by using loan guaranties.

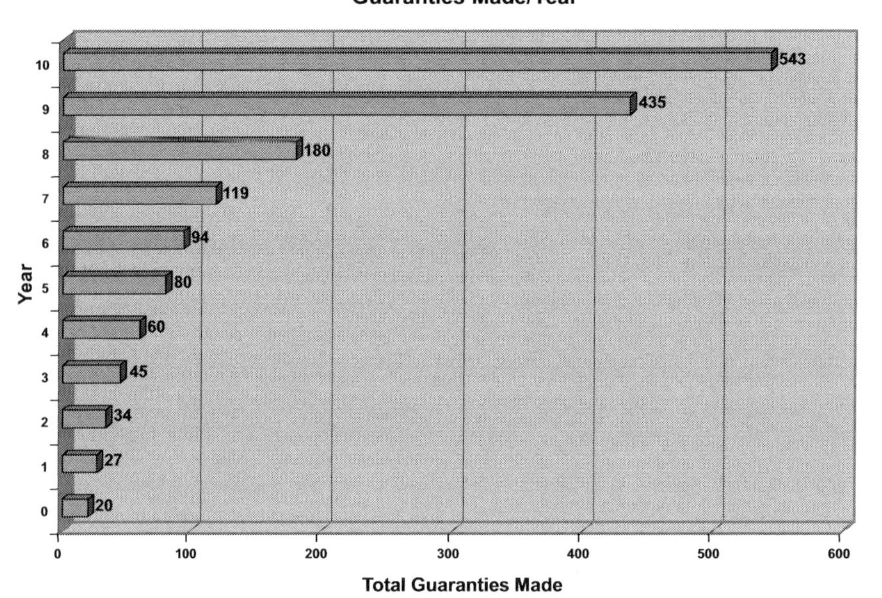

With loan guaranties, the NEF can fund a total of 543 projects over a ten year period! This amounts to 2,715% more projects than could have been funded by grants, 670% more projects than could have been funded with subsidized loans, and 374% more projects than could have been funded with a market rate loan program.

Loan guaranty programs are so efficient because they incorporate leverage. Leverage, in financial terminology, is the ability to increase the effect of the use of money. Here the money was the same as it was for grants: $100,000,000. Under the subsidized loan program, there was modest leverage because the funds were paid and could be re-loaned for another project. Thus, the same amount of money could be used twice. This is leverage.

In market rate loan program, there was additional leverage. This was because not only was the $100,000,000 of principal repaid, but 10% interest was also paid each year. So, this increased the leverage.

With loan guaranties an even greater leverage occurs because of the principle of insurance. A guaranty is the same as an insurance policy.

The principle of insurance is that not all events insured against will happen. In other words, if an insurance company insures the value of 1,000 automobiles valued at $5,000 each, it does not have to have $5,000,000 (i.e. enough money to pay for every car) in its reserve account. This is because from statistics it is known that out of 1,000 cars, only 2%–3% of them, or 20–30 cars, are likely to be destroyed in the course of a year. So, the insurance company knows that it will only have to pay out $100,000–$150,000. Thus it keeps about $250,000 to $500,000 in its reserve account to pay its losses.

A loan guaranty program works in the same manner. With $100,000,000, a fund such as the NEF should easily be able to guaranty over $1,000,000,000 of projects at any one time, because of the extreme unlikelihood that more than 10% of its projects would ever go into default at any one time. Thus, if the NEF were to guaranty commercial bank loans for $1,000,000,000 of projects, and 10% of them, or $100,000,000 of them were to default, the NEF could still make good on its guaranties by paying the banks holding the defaulted $100,000,000 held in their reserve account.

Thus, from the point of view of a government wishing to fund utility projects, the creation of a loan guaranty program would be the most efficient use of its funds.

FINANCIAL SIMULATION AND ANALYSIS OF USING THE FOUR FINANCING METHODS

Financial Simulation

Following financial simulation will be used to show the differences between:

- Market rate loans
- Grants
- Subsidized loans
- Loan guaranties

16

Assumptions and Conventions

Grant/Loan/Subsidy/Guaranty
Pro Formas

1. Government contributes $100,000,000 in year 0
2. All loans/guaranties made in each year on 31 December
3. Tenor of loans: 5 years, level principal method

17

Assumptions And Conventions

Grant/Loan/Subsidy/Guaranty
Pro Formas

4. For loan guaranties, estimated average project size: $5,000,000
5. Interest rate on subsidized loans: 0%
6. Interest rate on market rate loans: 10%
7. Interest Rate on Guaranty Fund: 5%

18

Assumptions And Conventions

8. Leverage ratios for loan guaranties:

# Of loans under guaranty:	**Coverage ratio:**
0-20	100%
20-30	90%
30-40	80%
40-50	70%
50-60	60%
60-70	50%
70-80	40%
80-90	30%
90-100	20%
100+	10%

19

Simulation Results: Grants

Year	0	1	2	3	4	5	6	7	8	9	10
Opening Balance	100	0	0	0	0	0	0	0	0	0	0
Interest Earned	0	0	0	0	0	0	0	0	0	0	0
Principal Repaid	0	0	0	0	0	0	0	0	0	0	0
Total Funds Available	100	0	0	0	0	0	0	0	0	0	0
New Grants Made	100	0	0	0	0	0	0	0	0	0	0
Ending Balance	0	0	0	0	0	0	0	0	0	0	0
Grants Made	20	0	0	0	0	0	0	0	0	0	0
Total Grants Made	20	20	20	20	20	20	20	20	20	20	20

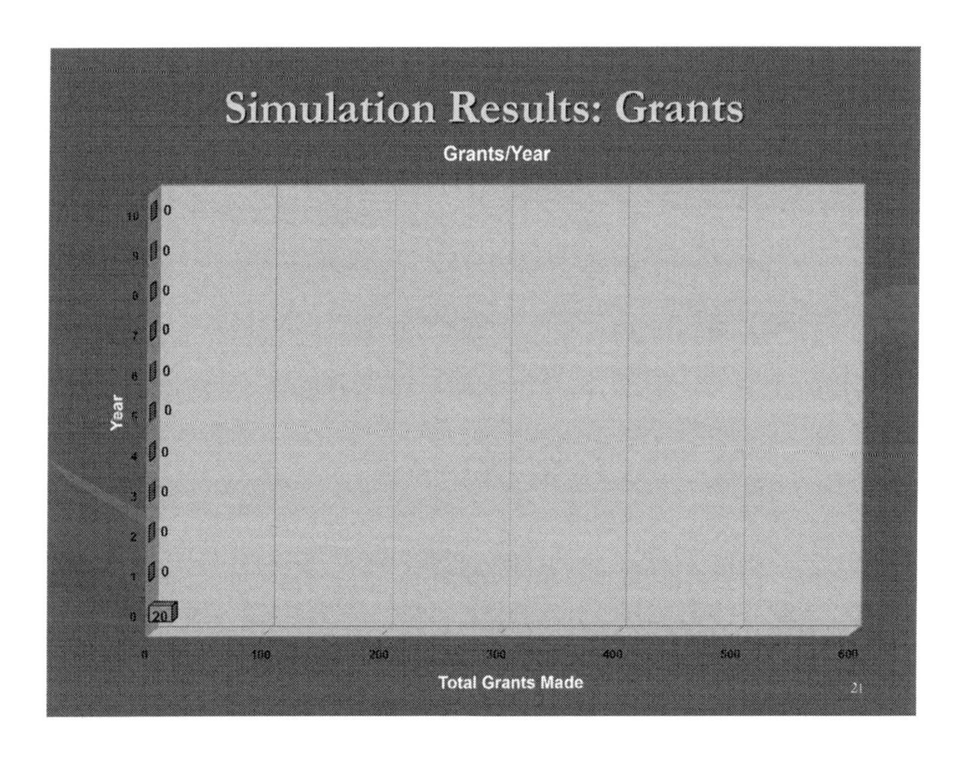

Simulation Results: Subsidized Loans

Year	0	1	2	3	4	5	6	7	8	9	10
Opening Balance	100	0	0	4	2	1	2	1	2	0	4
Interest Earned	0	0	0	0	0	0	0	0	0	0	0
Principal Repaid	0	20	24	28	34	41	29	31	33	34	33
Total Funds Available	100	20	24	32	36	42	31	32	35	34	37
New Loans Made	100	20	20	30	35	40	30	30	35	30	35
Ending Balance	0	0	4	2	1	2	1	2	0	4	2
Outstanding Loan Balance	100	100	96	98	99	98	99	98	100	96	98
Loans Made	20	4	4	6	7	8	6	6	7	6	7
Total Loans Made	20	24	28	34	41	49	55	61	68	74	81

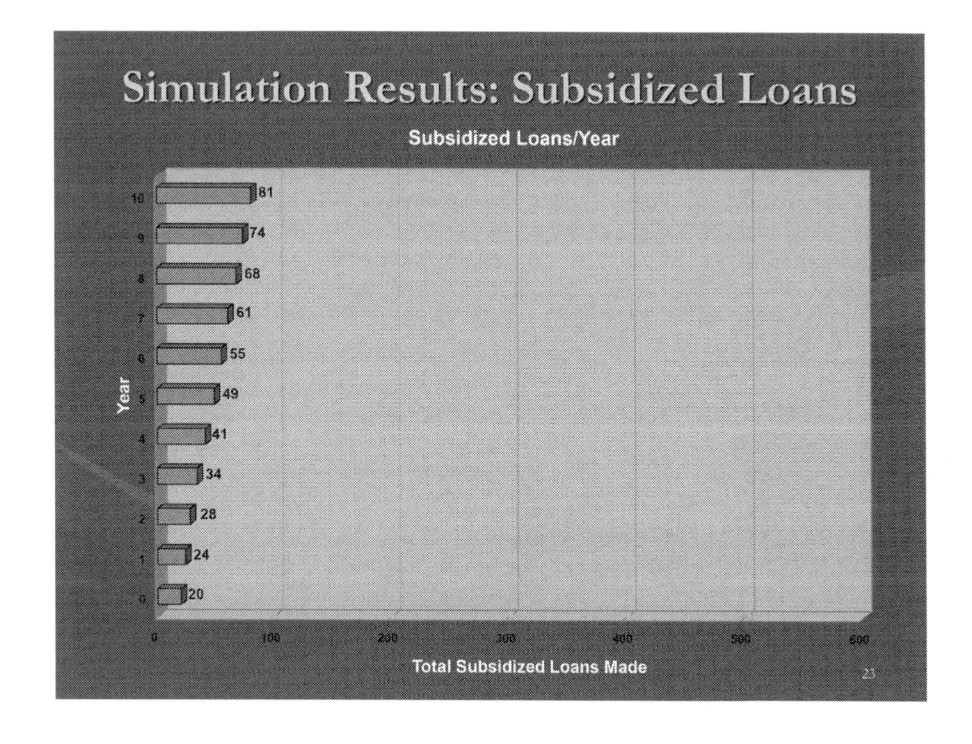

Simulation Results: Market Loans

Year	0	1	2	3	4	5	6	7	8	9	10
Opening Balance	100	0	0	2	1.9	2	4.4	1	1.5	0.7	3.9
Interest Earned	0.0	10	11	11.9	13.1	14.4	15.6	17.5	19.2	21.2	23
Principal Repaid	0.0	20	26	33	42	53	46	53	60	67	73
Total Funds Available	100	30	37	46.9	57	69.4	66	71.5	80.7	88.9	99.9
New Loans Made	100	30	35	45	55	65	65	70	80	85	95
Ending Balance	0.0	0.0	2	1.9	2	4.4	1	1.5	0.7	3.9	4.9
Outstanding Loan Balance	100	110	119	131	144	156	175	192	212	230	252
Loans Made	20	6	7	9	11	13	13	14	16	17	19
Total Loans Made	20	26	33	42	53	66	79	93	109	126	145

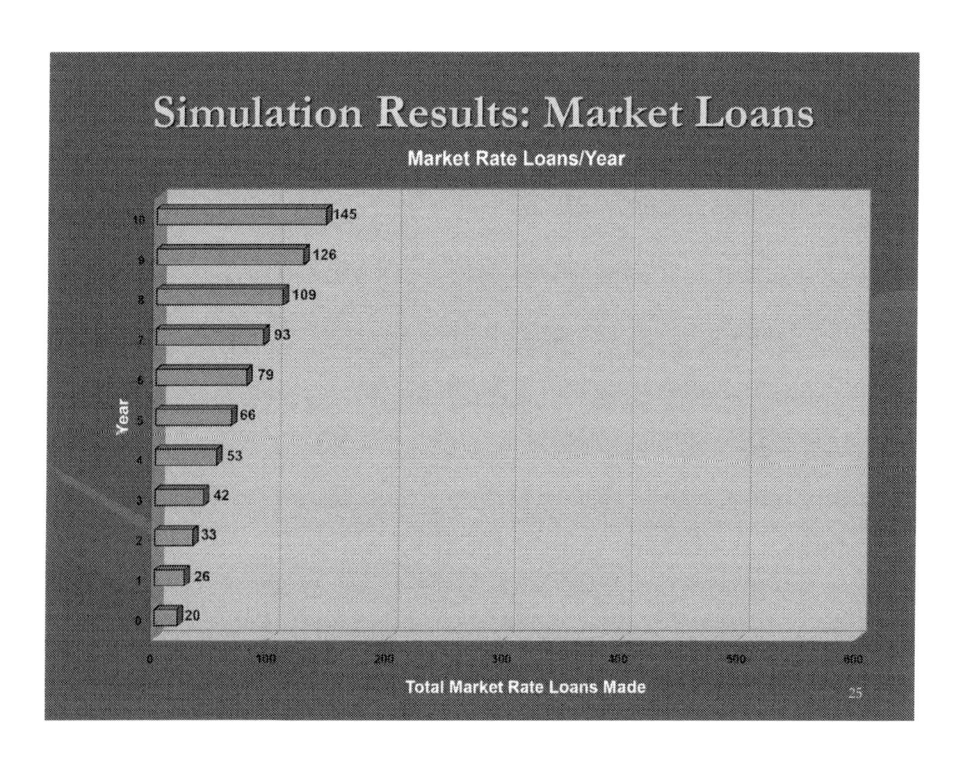

Simulation Results: Loan Guaranties

Year	0	1	2	3	4	5	6	7	8	9	10
Opening Fund Balance	100.0	100.0	105.5	111.4	117.5	124.1	131.2	138.9	147.0	155.8	166.2
Interest Earned	0.0	5.0	5.3	5.6	5.9	6.2	6.6	6.9	7.3	7.8	8.3
Guaranty Fees Earned	0.0	0.5	0.6	0.6	0.7	0.9	1.1	1.2	1.5	2.6	8.3
Closing Fund Balance	100.0	105.5	111.4	117.5	124.1	131.2	138.9	147.0	155.8	166.2	182.8
Leverage Ratio	100%	90%	90%	80%	70%	60%	60%	50%	30%	10%	10%
Maximum Guaranties	100	117	124	147	177	219	231	294	519	1662	1828
Begin Guaranties In Use	0	100	115	123	145	176	218	231	293	516	1659
Guaranties Released	0	20	27	33	44	58	57	63	82	132	372
Guaranties In Use	0	80	88	90	101	118	161	168	211	384	1287
New Guaranties Issued	100	35	35	55	75	100	70	125	305	1275	540
End Guaranties In Use	100	115	123	145	176	218	231	293	516	1659	1827
Guaranties Issued	20	7	7	11	15	20	14	25	61	255	108
Total Guaranties Issued	20	27	34	45	60	80	94	119	180	435	543

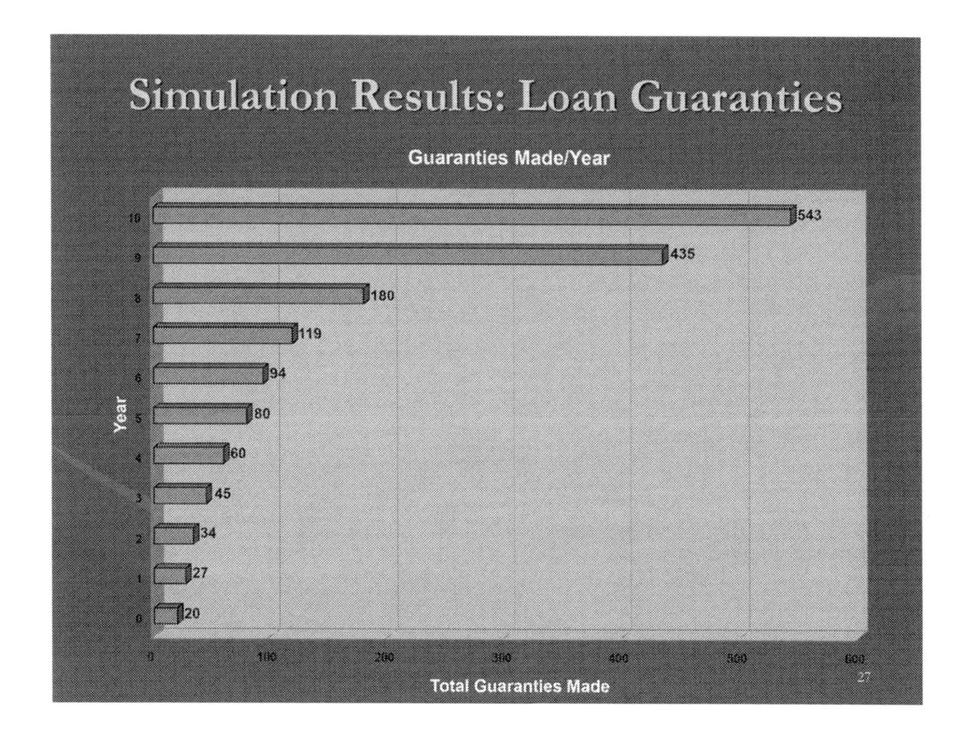

Analysis

Year	0	1	2	3	4	5	6	7	8	9	10
Total Market Rate Loans Made	20	26	33	42	53	66	79	93	109	126	145
Total Subsidized Loans Made	20	24	28	34	41	49	55	61	68	74	81
Total Guaranties Issued	20	27	34	45	60	80	94	119	180	435	543
Total Grants Made	20	20	20	20	20	20	20	20	20	20	20

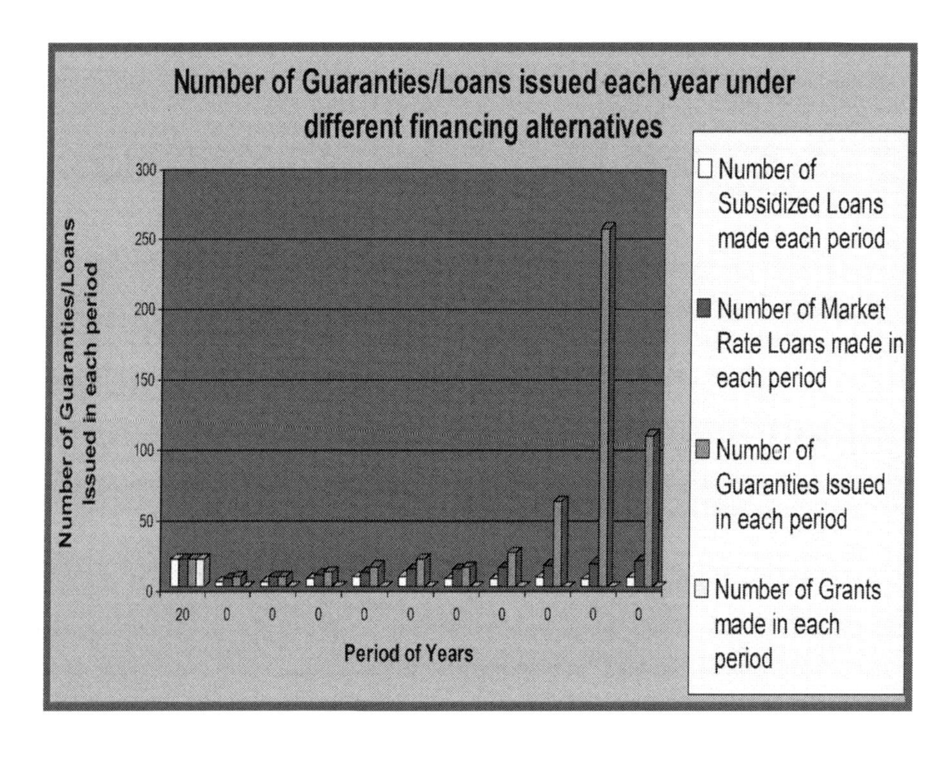

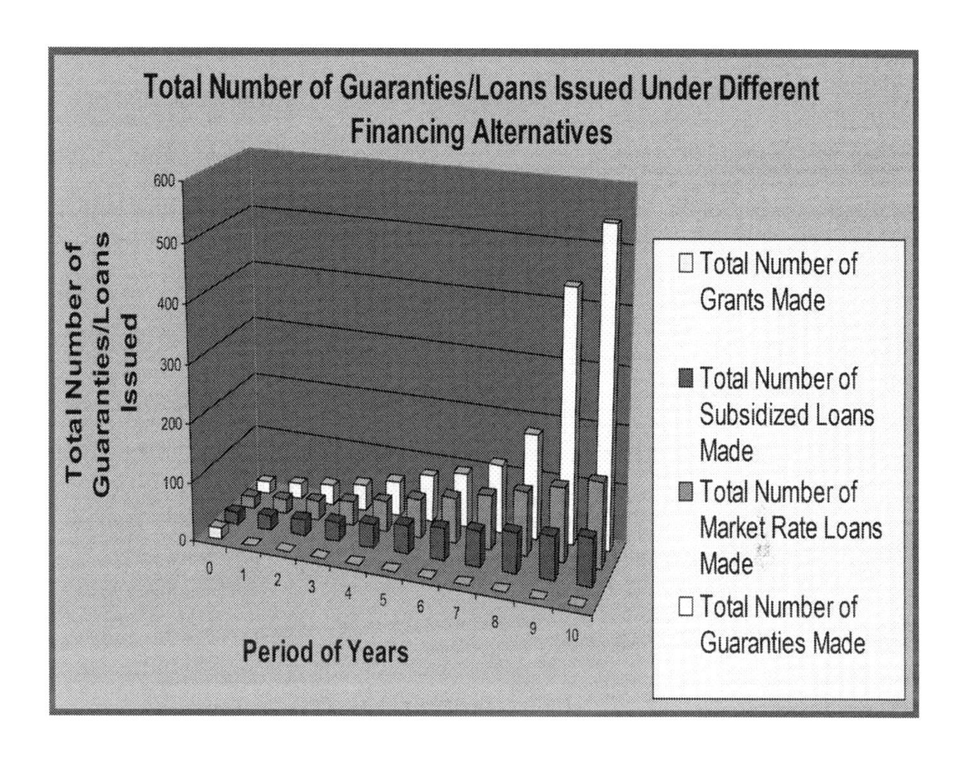
Total Number of Guaranties/Loans Issued Under Different Financing Alternatives
Total Number of Guaranties/Loans Issued
600
500
400
300
200
100
0
0
1
2
3
4
5
6
7
8
9
10
Period of Years
Total Number of Grants Made
Total Number of Subsidized Loans Made
Total Number of Market Rate Loans Made
Total Number of Guaranties Made

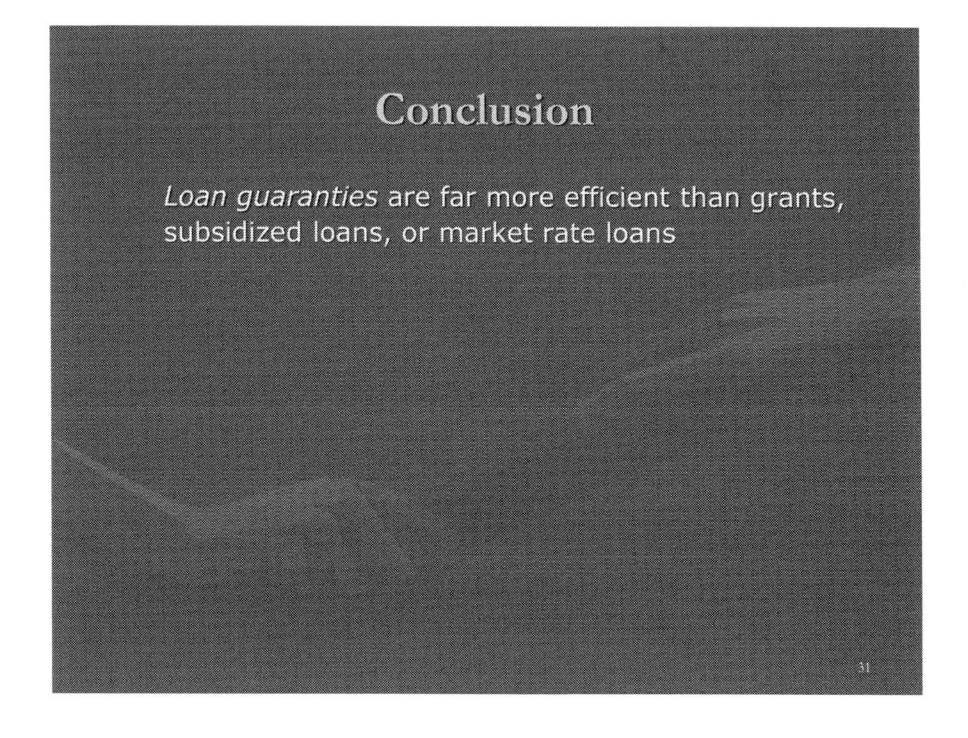
Conclusion
Loan guaranties are far more efficient than grants, subsidized loans, or market rate loans
31

Questions and Answers

Question #1:

Assume the existence of The International Sustainability Fund (ISF), a government agency that has seed money in the amount of $200,000,000. ISF managers decide to fund all of their sustainability projects with *grants*. Each project will cost $10,000,000. How many projects will ISF be able to fund?

Question #2:

The International Sustainability Fund (ISF) managers have decided to use the $200,000,000 seed money it has available to make *subsidized* loans to finance all of its sustainability projects. Loans are for a period of four years, repaid according to the level principal method. The interest earned on the subsidized loans is 0%. All money that is received by the ISF as repayment of principal is re-loaned in order to finance additional projects. Since all projects require $10,000,000, there may be some money (less than $10,000,000) rolled over to the next year. How many total subsidized loans can be made with the same $200,000,000 over an eight year period? Create a table to illustrate the annual loan(s) dispersed, the cumulative loans dispersed and the annual principal payment made to ISF from the borrower(s).

Question #3:

The International Sustainability Fund (ISF) managers have decided to use the $200,000,000 seed money it has available to make *market rate* loans to finance all of its sustainability projects. Loans are for a period of four years, repaid according to the level principal method. The current market rate is 5% for loans of $10,000,000 over a four year period. Both interest earned and principal repayment received by the ISF is re-loaned in order to finance additional projects. Since all projects require $10,000,000, there may be some money (less than $10,000,000) rolled over to the next year. How many total market rate loans can be made with the same $200,000,000 over an eight year period? Create a table to illustrate the annual loan(s) dispersed, the cumulative loans dispersed, the annual interest payment and the annual principal payment made to ISF from the borrower(s).

Question #4:

The International Infrastructure Development Fund (IIDF) managers have decided to use the $10,000,000 seed money it has available to make *subsidized* loans to finance all of its development projects. Loans are for a period of five years, repaid according to the level principal method. The interest earned on the subsidized loans is 0%. All money that is received by the IIDF as repayment of principal is re-loaned in order to finance additional projects. Since all projects require $2,500,000, there may be some money (less than $2,500,000) rolled over to the next year. How many total subsidized loans can be made with the same $10,000,000 over a five year period? Create a table to illustrate the annual loan(s) dispersed, the cumulative loans dispersed and the annual principal payment made to IIDF from the borrower(s).

Answer #1:

ISF will be able to fund 20 projects.

Answer #2:

Using the Level Principal Payment Method, in the case of a four year loan of $10,000,000 at 0% interest, the annual debt service payment would look like the following:

Annual Principal Payment = $10,000,000/4 = $2,500,000

Year	Annual Loan(s)	Cumulative Loans	Principal Payment
0	20	20	$0
1	5	25	$50,000,000
2	6	31	$62,500,000
3	8	39	$77,500,000
4	9	48	$97,500,000
5	7	55	$70,000,000
6	8	63	$75,000,000
7	8	71	$80,000,000
8	8	79	$80,000,000

The total amount of subsidized loans made over the eight year term is 79.

Answer #3:

Using the Level Principal Payment Method, in the case of a four year loan of $10,000,000 at 5% interest, the annual debt service payment would look like the following:

Annual Principal Payment = $10,000,000/4 = $2,500,000

Year	Annual Loan(s)	Cumulative Loans	Interest Payment	Principal Payment
0	20	20	$0	$0
1	6	26	$10,000,000	$50,000,000
2	7	33	$10,500,000	$65,000,000
3	9	42	$10,750,000	$82,500,000
4	12	54	$11,125,000	$105,000,000
5	10	64	$11,875,000	$85,000,000
6	10	74	$12,625,000	$95,000,000
7	12	86	$12,875,000	$102,500,000
8	12	98	$13,750,000	$110,000,000

The total amount of market rate loans made over the eight year term is 98.

Answer #4:

Using the Level Principal Payment Method, in the case of a five year loan of $2,500,000 at 0% interest, the annual debt service payment would look like the following:

Annual Principal Payment = $2,500,000/5 = $500,000

Year	Annual Loan(s)	Cumulative Loans	Principal Payment
0	20	20	$0
1	5	25	$50,000,000
2	6	31	$62,500,000
3	8	39	$77,500,000
4	9	48	$97,500,000
5	7	55	$70,000,000
6	8	63	$75,000,000
7	8	71	$80,000,000
8	8	79	$80,000,000

The total amount of subsidized loans made over the five year term is nine.

8 Sources of Funds

INTRODUCTION

The first six chapters in this series constitute the first six steps in developing the financial information necessary for undertaking projects where income is associated with the delivery of certain utility services, whether the utility is publicly or privately owned. This includes projects in the following utility sectors: water, wastewater, solid waste, and certain types of energy efficiency projects. Those six chapters are written from the perspective of a utility executive who needs a project to improve his system and is interested in learning the financial techniques necessary to obtain the money to carry out the project.

After introducing the concept of project finance (Chapter 1), the first step was the presentation of methods for measuring income (Chapter 2 – Measuring Income). Methods for maximizing income were presented in the succeeding chapter (Chapter 3 – Maximizing Cash Available for Debt Service). It was then emphasized that the maximization of income was of paramount importance because if income exceeded operating expenses, needed projects could be undertaken to improve utility systems by having the utility incur debt. If they had excess income, they would be free of a pernicious reliance on grants. Excess income can be used to pay annual debt service. The next chapter, therefore, was devoted to the explanation of the basic concepts of loans and debt (Chapter 4 – Loan Basics).

Once the basic concepts of borrowing are understood, the next step is to be able to evaluate various types of loans in order to determine which available loan is best for a particular utility. Techniques for evaluating loans, therefore, were presented in the following chapter (Chapter 5 – Project Valuation).

The next chapter used all of the information presented in the preceding five chapters to illustrate which projects were financially feasible for a utility to undertake. It demonstrated both how the largest possible project could be undertaken at the lowest possible cost and how a utility's excess operating income could be used to improve the system (Chapter 6 – Financial Feasibility).

In the last chapter (Chapter 7 – Alternate Finance Sources), however, the perspective shifts. Financial concepts were no longer presented from the utility executive's perspective. Rather, in this chapter, financial concepts were presented from the perspective of a government official whose responsibility it is to manage or administer a program to provide funds for municipal utility services. In other words, in the first six chapters, project finance was discussed from the perspective of one needing funds (for a particular project). In the seventh chapter, the concepts of project finance were discussed from the perspective of a government official who has funds, or whose job it is to seek funds for utility projects in his country. It is this official's responsibility to provide the best quality of utility services to the largest amount of people by providing funds to utility operators for projects, which improve that utility's services to its users.

In this eighth chapter, the same perspective is maintained as in the previous chapter. In other words, this chapter is written for a government official whose responsibility is to manage a program to finance projects for utilities. In this chapter, sources of external funding for such programs are examined and compared with a view to the most likely and most reliable sources of funding for utility projects.

The United Nations has estimated that the cost of reaching the Millennium Development Goal of halving the number of people without access to clean water or basic sanitation by 2015 is an additional 100 billion USD per year. There are four sources of funding for environmental projects. This chapter will examine these four sources of capital.

CHAPTER CONTENTS

This chapter will discuss the following four topics as sources of funding for environmental utility and other utility type of funding:

- National Budgets
- International Financial Institutions
- Donors
- Private Capital

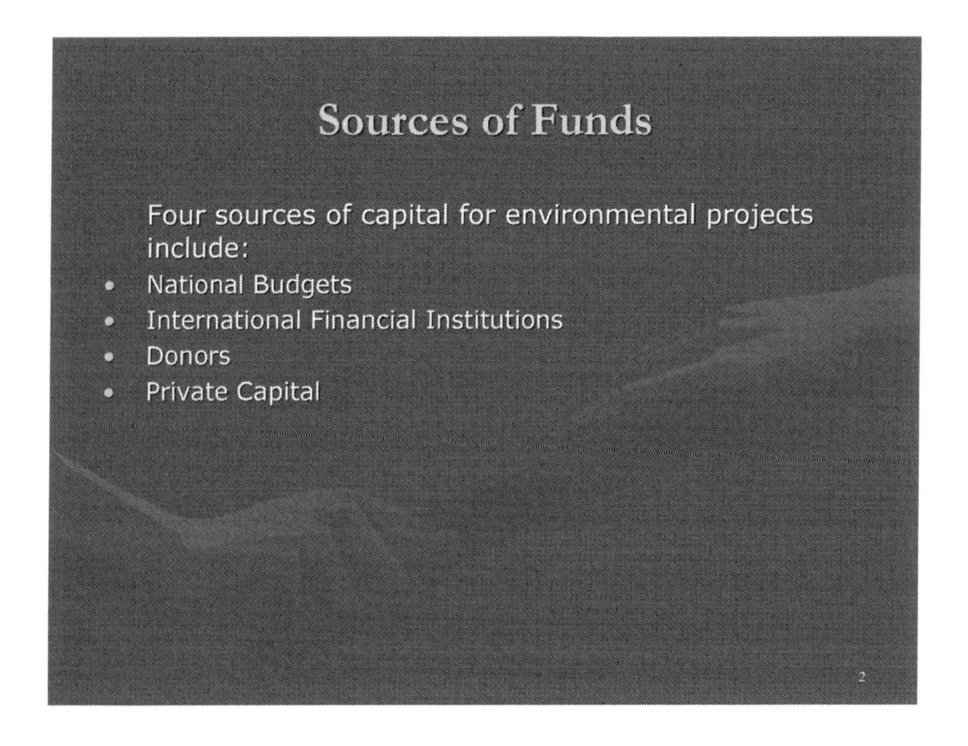

NATIONAL BUDGETS

In every country, the environment must compete for scarce central government funds with such high priority issues as national defense, food, public health, education, housing and economic development.

National Budgets

In every country, the environment must compete for scarce central government funds with such high priority issues as national defense, food, public health, education, housing and economic development.

INTERNATIONAL FINANCIAL INSTITUTIONS

International Financial Institutions

The World Bank and all of the regional development banks **combined** do not have the financial resources needed to pay for clean water and clean air across the globe.

4

The World Bank, whose formal title is the International Bank for Reconstruction and Development (IBRD), and all of the regional development banks combined do not have the financial resources to pay for clean water and clean air across the globe. The World Bank is actually two institutions: IBRD itself, and the International Development Association (IDA). The World Bank also has two important affiliates, the Multilateral Investment Guaranty Authority (MIGA) and the International Finance Corporation (IFC). Together, all of these institutions are referred to as the World Bank Group.

There are four major regional international development banks: the InterAmerican Development Bank (IADB) serving North and South America; the Asian Development Bank (ADB), the African Development Bank (AfDB); and the European Bank for Reconstruction and Development (EBRD), which serves the countries of the former Soviet Union. There are also several other specialized international development banks such as the Islamic Development Bank and the Caribbean Development Bank. There are also some sector-specific institutions such as the Global Environment Facility (GEF) and the Nordic Environmental Finance Corporation (NEFCO).

The United Nations has estimated that all of the international development banks combined lend approximately $1 billion per year for water and wastewater projects.

One of the principal barriers to increasing this sum is the recipient countries themselves. The international development banks (with two exceptions discussed below) make loans to sovereign governments. Sometimes the recipient governments themselves request loans for water projects. Other times the international development banks encourage governments to borrow funds for water and wastewater projects based on studies the banks have made of the recipient countries infrastructure and public health needs. In any event, none of the projects funded by the international development banks may be financially sustainable. This is because the international development banks lend directly to the recipient government which must repay the loans – neither wholly nor partly from project revenues – but rather from general central government revenues. The recipient government may then on lend the money to a local government or a water/wastewater utility, or they may furnish the money to the project in the form of a grant. In either case, the central government is obligated to repay the loan to the international development bank. If project revenues are insufficient to repay the loan, then the central government must tap other sources of revenues to repay the international development bank.

There are two exceptions to the sovereign lending rule. The EBRD makes sub-sovereign loans to municipal governments without requiring the guaranty of the central government.

The IFC also makes loans to sub-sovereign units of government through its "Municipal Fund". Although the IFC charter restricts its lending activity to the private sector, the "Municipal Fund" will make loans to public sector borrowers if such funding serves to facilitate private sector lending. Thus, the IFC could, for example, make a loan to a sub-sovereign unit of government to fund a bond fund reserve for a municipal bond bank under the theory that the bond bank will issue bonds that will be purchased by pension funds, life insurance companies, and other institutions in the private sector.

International Financial Institutions

- United Nations estimates all international development banks combined lend approximately $1 billion/year for water and wastewater projects
- Estimated cost of reaching the Millennium Development Goal of halving the number of people without access to clean water and sanitation by 2015 will be an additional **$100 billion/year**

5

DONORS

Donor countries can only provide a small portion of the funds needed for a clean environment. At present, they furnish about $3 billion a year for water and wastewater. In addition, the environment must compete for donors' scarce funds with other issues such as food, public health, housing, and education.

Major donors are the G-8 countries: the United States, Canada, Great Britain, France, Germany, Japan, Italy, Spain, and the Netherlands. The European Union is now a major donor in its own right. And, as its trade expands across the globe, China is becoming a major donor as well. Venezuela has also begun to make grants to other countries in Latin America and the Caribbean.

THE PRIVATE SECTOR

Only private capital can provide the vast amounts of money needed to provide clean water and clean air to the billions of people on this planet. The world economy is approximately 35 trillion USD. As such, the $100 billion needed to reach the Millennium Development Goal for safe drinking water and basic sanitation represents about 3/10ths of 1%.

Private Capital

- Only private capital can provide the vast amounts of money needed to provide clean water and clean air to the billions of people on this planet
- The world economy is approxiamately $35 trillion
- The $100 billion needed to reach the Millennium Development Goal for safe drinking water and sanitation represents about 3/10ths of 1% of the above amount

7

How to Access Private Capital

How to Access Private Capital

8

Private capital requires high returns for high risk investments, and low returns on low risk investments. Environmental projects cannot pay high returns without decreasing the number of projects being completed (given small and inflexible budgets). Since environmental projects can only make low payments on loans, they must be a low risk investment for lenders.

Private Capital Requires

- High returns on high risks
- Low returns on low risks

9

High Return on High Risks

- Environmental projects cannot pay high returns
- High payments for environmental projects result in fewer projects being built

10

Low Return on Low Risks

Environmental projects must pay low returns, therefore, environmental projects must be **low risk**.

11

TYPES OF RISK

Private Capital Recognizes 2 Types of Risk

- Liquidity Risk
- Credit Risk

12

When assessing investment options, private investors evaluate two types of risk: credit and liquidity.

> *Liquidity Risk:* the risk that the investor will not be able to sell his investment prior to maturity; an investment with no resale value is illiquid.

Liquidity Risk: A liquid market has a large number of similar investment opportunities. To be liquid, environmental finance programs must be as broad based as possible – nationwide, not regional in scope. A liquid market is also permanent; similar investments must be available every year. Environmental finance programs cannot depend on the annual acts of a national legislature. To be permanent, an environmental finance program must be self-sustaining.

Liquidity Risk

- *Liquidity Risk*: the risk that the investor will not be able to sell the investment prior to maturity
- An investment with no resale value is illiquid
- Characteristics of a liquid market:
  - Large number of similar investment opportunities
  - Permanent, similar investments available every year
- To be liquid, an environmental finance program must be broad based and self-sustaining

13

Self-Sustaining Environmental Finance Programs: a program that can provide funding for environmental projects indefinitely without having to rely on governmental funding; revenues primarily come from private capital investments and loan repayments from environmental projects.

A broad-based, self-sustaining environmental finance system – where all projects can be funded without reliance on an annual governmental budget allocation – will be liquid and, as such, will attract private capital investments.

Credit Risk: The second type of risk that the private sector must deal with is credit risk.

Credit Risk

- *Credit Risk*: the risk that the investor will not be repaid or, in other words, the risk of non-payment.
- Based on CADS of utility
- Must convey favorable CADS to private sector

14

Much has been written in the preceding seven chapters about the need to maximize water/wastewater system revenues and minimize expenses. The end-goal of these efforts is to produce excess cash, which can be used to pay debts. Thus Cash Available for Debt Service is the core concept in assessing credit risk. Private investors must ask if the water/wastewater utility borrower is creditworthy. In other words, can the utility consistently and reliably produce the amount of Cash Available for Debt Service necessary to repay a particular loan. Once a loan is made, the investor must take the risk that the utility will continue to produce enough Cash Available for Debt Service to make periodic payments until the loan is fully repaid.

Not only must potential water/wastewater utility borrowers produce a substantial and reliable stream of Cash Available for Debt Service; but they must also convey this information to the private sector investors in order to convince them to make loans or purchase bonds. In this regard, there are two fundamental principles that must be observed in order to attract private sector investors.

The first is that financial records and other financial information must be produced in accord with internationally accepted accounting standards. No private sector investor will lend to a potential borrower with incomplete, inconsistent, or idiosyncratic financial records. International accounting norms have been firmly established for all sectors of the economy, including water and wastewater utilities. These standards and norms must be scrupulously observed if utilities ever intend to borrow from private sector investors.

The second principal is transparency. A utility must produce monthly cash reports, quarterly financial statements, and annual financial reports that are audited by an independent third party professional accountant. Furthermore, these records – at least the quarterly statements and annual financial reports – must be readily available to investors. They must, of course, be prepared in accord with internationally accepted accounting standards. They must either be published and available to the public; or they must, at least, be available to investors during all normal business hours.

Transparency means that a utility's financial performance must be open to the exacting scrutiny of professional private sector investors. Anything less than full transparency, and any format other than internationally accepted accounting standards, will surely discourage any private sector investor from lending to any water/wastewater utility borrower.

Conclusion

In conclusion, there are few sources of grant funding. Donors make grants. International development banks can make a limited amount of grants from trust funds that they may administer.

Loan funds, on the other hand, are plentiful. Not from international development banks nor from donors, but rather from the private sector. And, the only way to attract private sector investment is to have Cash Available for Debt Service and to address the two critical issues that face the private sector: liquidity risk and credit risk. Liquidity risk can be addressed by the creation of permanent, broad-based programs that are not dependent on government funds for their operation. Credit risk – once a utility has accumulated a steady stream of Cash Available for Debt Service by adopting internationally accepted accounting standards for all of its financial reporting and by making its financial books and records available to the public in a fully transparent manner.

The only way for nations to access the private capital market is to create permanent, nationwide, self-sustaining systems for financing environmental projects.

Attracting Private Investors

Two fundamental principles used to attract private sector investors:

- Financial information produced in accord with internationally accepted accounting standards
- Transparency of financial statements and reports, readily available to investors

15

Questions and Answers

Question #1:

What are the four sources of funding for environmental projects?

Question #2:

The United Nations estimates all international development institutions combined lend approximately $1 billion/year for water and wastewater projects. What is the estimated additional cost per year in US dollars for reaching the Millennium Development Goal of halving the number of people without access to clean water or basic sanitation by the year 2015?

Question #3:

Given small and inflexible budgets, what type of loan investment, low or high risk, will an environmental project be considered to a lender? Why?

Question #4:

When assessing investment options, what two types of risk do private investors evaluate?

Question #5:

Would a Self-Sustaining Environmental Finance Program be an example of a liquidity or credit risk?

Question #6:

What are two fundamental principles that must be observed in order to attract private sector investors?

Answer #1:

National budgets, international financial institutions, donors, and private capital.

Answer #2:

An additional $100 billion/year.

Answer #3:

Low risk, due to the fact that environmental projects cannot pay high returns without decreasing the number of projects being completed.

Answer #4:

Liquidity risk and credit risk.

Answer #5:

Liquid.

Answer #6:

Financial records and other financial information must be produced in accord with internationally accepted accounting standards. A utility must be transparent by producing monthly cash reports, quarterly financial statements, and annual financial reports that are audited by an independent third party professional accountant and readily available to investors.

9 Cost/Benefit Analysis

INTRODUCTION

The first six chapters in this book constitute the first six steps in developing the financial information necessary for undertaking projects where income is earned from the delivery of certain utility services, whether the utility is publicly or privately owned. The focus of these chapters will be on water and wastewater utilities; but the same principles apply to other similar types of utilities such as those that deliver power, heat, or solid waste services. The six chapters are written from the perspective of a utility executive who must undertake a project to improve his system and is interested in learning the financial techniques necessary to obtain the money needed for the project.

After introducing the concept of project finance (Chapter 1), the first step is the presentation of methods for measuring income (Chapter 2 – Measuring Income). All money is not the same. Not all income can be used for project debt. Chapter 1 defines what income can be used for making debt service payments and how it can be estimated.

Methods for maximizing income are presented in the succeeding chapter (Chapter 3 – Maximizing Cash Available for Debt Service). It is emphasized that the maximization of income is of paramount importance because if income exceeds operating expenses, needed projects can be undertaken to improve utility systems by having the utility incur debt. If utilities have excess income, they can be free of the pernicious reliance on grants. Excess income can be used to pay annual debt service. The next chapter, therefore, is devoted to the explanation of the basic concepts of loans and debt (Chapter 4 – Loan Basics).

Once the basic concepts of borrowing are understood, the next step is to be able to evaluate various types of loans in order to determine which available loan is best for a particular utility. Techniques for evaluating loans, therefore, are presented in the following chapter (Chapter 5 – Project Valuation).

Chapter 6 – Financial Feasibility, uses all of the information presented in the preceding five chapters to illustrate which projects are financially feasible for a utility to undertake. It demonstrates both how the largest possible project could be undertaken at the lowest possible cost and how a utility's excess operating income can be used to improve the system.

In Chapter 7 – Alternate Finance Sources, the perspective shifts. Financial concepts are no longer presented from the utility executive's perspective. Rather, in this chapter, financial concepts are presented from the perspective of a government official whose responsibility it is to manage or administer a program to provide funds for municipal utility services – an environmental infrastructure finance program.

In other words, in the first six chapters, project finance is discussed from the perspective of one needing funds (for a particular project). In the seventh chapter, the concepts of project finance are discussed from the perspective of a government official who has funds, or whose job it is to seek funds for utility projects in his country. It is this official's responsibility to provide the best quality of utility services to the largest number of people by providing funds to utility so that they can provide high quality services to their customers.

In Chapter 8, Sources of Funds, the same perspective was maintained as in the previous chapter. In other words, it was written for a government official whose responsibility is to manage a finance program for utilities. There, sources of external funding for such programs are examined and compared with a view to the most likely and most reliable sources of funding for utility projects.

These chapters, Chapter 9 – Cost/Benefit Analysis, along with Chapter 7 – Alternate Finance Options, are the two most important elements in this series of training materials.

CHAPTER CONTENTS

- Rubric for Grantmakers
- Matrix Requirements

RUBRIC FOR GRANTMAKERS

In Chapter 7, the point was made that the most efficient use of funds for environmental infrastructure is loan guaranties. The point was also made that grant funds are the least efficient use of funds. Grant monies – once disbursed – are gone, forever.

Notwithstanding the fact that grants are extremely inefficient; they are nonetheless necessary in most circumstances. That is because many utilities cannot set their tariff rates high enough to pay 100% of the debt service on needed projects. When a utility cannot pay the full debt service on a loan to cover 100% of the cost of a project, then the amount of the loan must be reduced by a grant. This is called a "buy-down". The grant buys-down the project cost to the point where the utility has enough Cash Available for Debt Service to repay a loan.

Now, the overwhelmingly important question that faces government financial policy makers is: Which project receives a grant? And, how much should the grant be?

Introduction

To review:

- Grants are extremely inefficient, but necessary in most circumstances
- For example, many utilities cannot set their tariff rates high enough to pay 100% of the debt service needed on projects
- A grant is needed to reduce the project cost to the point where the utility has enough CADS to repay the loan, this is called a *buy-down*

2

The answer to these questions is deceptively simple: “the project which requires the smallest amount of grant funds that does the most environmental good for the largest number of people”.

This answer can be embodied in a rubric, which can be used as a rule for grantmakers.

For drinking water projects, the rubric can be stated as follows:

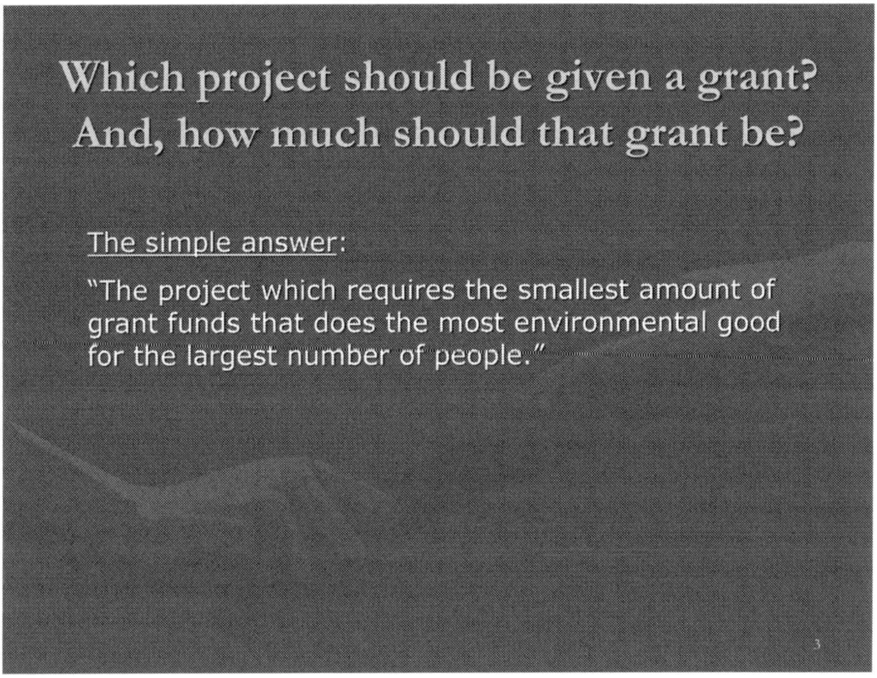

Which project should be given a grant?
And, how much should that grant be?

The simple answer:

"The project which requires the smallest amount of grant funds that does the most environmental good for the largest number of people."

3

"The highest priority for the use of limited grant funds is for projects which:

1) reduce or eliminate the most severe waterborne disease
2) for the greatest number of people, while
3) requiring the smallest amount of grant."

Drinking Water Rubric for Grantmakers

The highest priority for the use of limited grant funds is for **drinking water** projects which:

1. reduce or eliminate the most severe waterborne disease

2. for the greatest number of people, while

3. requiring the smallest amount of grant.

For wastewater projects, the rubric would be very similar, as follows:

"The highest priority for the use of limited grant funds is for projects which:

1) reduce or eliminate the largest quantity of
2) the most toxic pollutants, while
3) requiring the smallest amount of grant."

Wastewater Rubric for Grantmakers

The highest priority for the use of limited grant funds is for **wastewater** projects which:

1. reduce or eliminate the largest quantity of
2. the most toxic pollutants, while
3. requiring the smallest amount of grant.

5

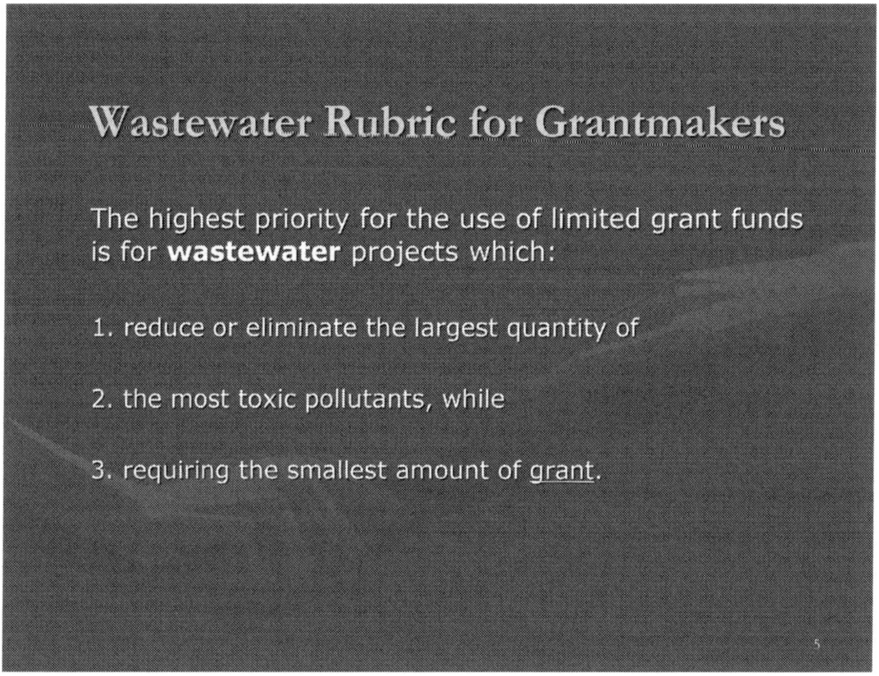

(Henceforth, only the drinking water rubric will be used in examples.)

As is apparent, the above rubrics relate the cost (in terms of grant amount) to the benefits it produces. In short, to effectuate this rubric, a government environmental finance program that makes grants must: 1) perform a cost/benefit analysis, and, 2) prioritize the results of that analysis. To achieve these two results, government environmental finance programs must create two matrices to deal with the data necessary for the cost/benefit analysis.

Note that the rubrics above specifically say grants, not loans, or funding, or anything else. Just grants. This is for a very specific reason. As stated previously, grant funds are very precious. This is because once a grant is made, the money is gone; unlike loans, where, once the loan is made, it begins to be repaid (with interest!) and the repayments can be recycled into new loans.

Rubric for Grantmakers (cont.)

A government environmental finance program that makes grants must:

1. perform a cost/benefit analysis, and

2. prioritize the results of that analysis

Note: The above rubric applies specifically to grants, due to the fact that once a grant is made, the money is gone. Loans do not apply since they can be repaid and recycled into new loans.

6

MATRIX REQUIREMENTS

Basic Matrix Requirements

Based on the previous drinking water rubric, the creation of two matrices will follow, along with a unique form of cost/benefit analysis. The following requirements will assist in the creation of these matrices:

Each country considered should have:

- A means of identifying and quantifying outbreaks of waterborne diseases
  - Chemical and bacteriological water testing
  - Anecdotal reporting, where scientific testing is unavailable
- Significant population numbers

Note: Henceforth, all examples will center on the **drinking water** rubric.

7

Note that the rubrics are deceptively simple. They imply the creation of a matrix – in fact two matrices – and a unique form of cost/benefit analysis. In doing so, they have requirements which have great value in and of themselves.

These requirements are:

1) Each country should have, or otherwise develop, a means of identifying and quantifying outbreaks of waterborne diseases. This can either be done through the reporting of incidences of diseases themselves, or through chemical and bacteriological water testing. In countries where there are no water tests or where the water tests are inadequate, the reporting of disease is the only way to measure water problems. Even where such reporting is unscientific and anecdotal, it can still be used. In the absence of numbers, anecdotes work. In the land of the blind, the one-eyed man is king. However, countries should still strive to build efficient and science-based systems for determining water quality.
2) Each country should have good population numbers. This seems obvious, but not necessarily so in countries where budget subventions to local governments are population based. Here, all the young people who flee to the cities to find jobs are still kept on the record books. This way, the village doesn't lose the aid payments allotted to its missing citizens.

The first matrix is then constructed, consisting of: a) the ranking extent of waterborne diseases by severity, b) the ranking of them by occurrence, and, c) the populations in areas of occurrence. This matrix then defines the first two elements of the rubric: "places where the greatest number of people suffer most frequently from the most severe waterborne diseases". This is the "benefit" matrix, which is the "benefit" side of the cost/benefit analysis.

"Benefit" Matrix

The first matrix constructed will define the **"benefit"** side of the cost/benefit analysis and will consist of the following elements:

- Ranking extent of waterborne diseases by severity
- Ranking of waterborne diseases by occurrence
- Populations in areas of occurrence

From the above matrix, the first two elements of the drinking water rubric can be defined:

Places where the greatest number of people suffer most frequently from the most severe waterborne diseases

8

The second matrix is the cost side of the analysis. It is more complicated, but its elements can easily be understood. They involve:

"Cost" Matrix

The second matrix constructed will define the **"cost"** side of the cost/benefit analysis and will consist of the following elements:

- Restructuring of tariffs so that they are consumption-based and involve incremental block tariffs (IBTs) (Mod. X)
- Restructuring of subsidies from general, supply-based subsidies to targeted, demand-based subsidies that serve only the truly needy (Mod. XI)
- Adoption of tariffs that reach or exceed full-cost recovery
- Modernization of billing and collection procedures to get collections over 95% (Mod. III)
- Identification of operational cost saving measures (Mod. III)
- Determination of the amount by which utility revenues exceed expenses (Mod. III), this number becomes CADS

9

1) Restructuring of tariffs so that they are consumption-based and involve incremental block tariffs (IBTs). (Cf. Chapter 10 – Tariff Design)
2) Restructuring of subsidies from general, supply-based subsidies to targeted, demand-based subsidies that serve only the truly needy. (Cf. Chapter 11 – Subsidies)
3) The adoption of tariffs that reach or exceed full-cost recovery.
4) The modernization of billing and collection procedures to get collections over 95%. (Cf. Chapter 3 – Maximizing Cash Available for Debt Service.)
5) The identification of operational cost saving measures. If they involve changes in operations, they should be effected. If they require capital investment (e.g., modern pumps to save energy), then they should be included in project costs and the savings reflected in cost projections. (Cf. Chapter 3 – Maximizing Cash Available for Debt Service.)
6) Based on the income maximization steps (Items 1–5, above), a determination of the amount by which utility revenues exceed expenses. This number becomes "Cash Available for Debt Service" for the project. (Cf. Chapter 3 – Maximizing Cash Available for Debt Service.)
7) Determination of the cost of the project.
8) Determination of the annual debt service by analyzing the rates and terms of financing available for the utility's project.
9) Determination of how much debt can be supported by the utility's Cash Available for Debt Service.
10) Determination of the amount of grant necessary to complete the project by subtracting the amount of debt which the utility's Cash Available for Debt Service can support from the total project cost. Assuming the utility has maximized its Cash Available for Debt Service, this number is, then, the smallest amount of grant necessary to undertake the project.

"Cost" Matrix

Cost matrix elements (cont.):

- Determination of the cost of the project
- Determination of the annual debt service by analyzing the rates and terms of financing available for the utility's project
- Determination of how much debt can be supported by the utility's CADS
- Determination of the amount of grant necessary to complete the project by subtracting the amount of debt the utility's CADS can support from the total project cost, this number is *the smallest amount of grant necessary to undertake the project* (the final element of the drinking water rubric is now defined)

10

11) Prioritization of all projects by grant amount from the smallest grant necessary to the largest. This is the second matrix.
12) The correlation of the two matrices, which will indicate where the "most severe waterborne disease affecting the greatest number of people can be reduced or eliminated with the smallest amount of grant funding". This is the total cost/benefit analysis.

Total Cost/Benefit Analysis

Now that the cost/benefit matrices have been constructed, the cost/benefit analysis can be performed:

- Prioritize all projects by grant amount from the smallest grant necessary to the largest
- Correlate the two matrices to indicate where the *"most severe waterborne disease affecting the greatest number of people can be reduced or eliminated with the smallest amount of grant funding."*

For government officials who are responsible for safe drinking water and for the abatement of water pollution, and who provide or allocate funds for water and wastewater infrastructure projects to deal with these problems, this type of analysis accomplishes two major goals. First, it assures the people and their government that the most severe sources of waterborne contamination (or pollution) are addressed. Second, it also assures the people and their government that their money is put to the most efficient use in attacking these types of problems.

Conclusion

Total cost/benefit analysis accomplishes two major goals:

Assures the people and their government that:

1. the most severe sources of waterborne contamination (or pollution) are addressed
2. their money is put to the most efficient use in attacking these types of problems

12

QUESTIONS AND ANSWERS

Question #1:

What is it referred to when a utility cannot pay the full debt service on a loan to cover 100% of the project cost and, therefore, the amount of the loan must be reduced by a grant?

Question #2:

The highest priority for the use of limited grant funds is for wastewater projects which:

1. reduce or eliminate the largest quantity of
2. the most toxic pollutants, while
3. requiring the largest amount of grant

Which statement(s) is/are true?

A) 1 and 3
B) 2 and 3
C) 2 only
D) 1 and 2
E) 1, 2, and 3

Question #3:

Prior to the creation of "cost/benefit" matrices and the application of Cost/Benefit Analysis in order to prioritize which water/wastewater projects should receive grant funds, there are two country-specific requirements that must be considered.

What are these two requirements?

Question #4:

The "benefit" matrix addresses the following elements of the drinking water rubric:

1. reduce or eliminate the most severe waterborne disease
2. for the greatest number of people, while
3. requiring the smallest amount of grant

Which statement(s) is/are true?

A) 1 and 2
B) 1 and 3
C) 2 and 3
D) 2 only
E) None of the statements true

Question #5:

Which element(s) is/are not involved in the "cost" side of the Cost/Benefit Analysis?

A) Restructuring of subsidies from general, supply-based subsidies to targeted, demand-based subsidies that serve only the truly needy.
B) Ranking extent of waterborne diseases by severity.
C) Determination of the annual debt service by analyzing the rates and terms of financing available for the utility's project.
D) Determination of the cost of the project.
E) Determination of how much debt can be supported by the utility's Cash Available for Debt Service.

Answer #1:

Buy-down.

Answer #2:

D) 1 and 2. Statement 3 is false. It should read as follows: The highest priority for the use of limited grant funds is for wastewater projects which require the smallest amount of grant.

Answer #3:

Each country considered should have:

1) a means of identifying and quantifying outbreaks of waterborne diseases; and,
2) significant population numbers.

Answer #4:

A) 1 and 2. Statement 3 is defined by the "cost" matrix.

Answer #5:

B) Ranking extent of waterborne diseases by severity.

10 Tariff Design

INTRODUCTION

This is the tenth in a series of eleven teaching units on the financing of environmental projects.

The first six units in this series constitute the first six steps in developing the financial information necessary for undertaking projects where income is associated with the delivery of certain utility services, whether the utility is publicly or privately owned. This includes projects in the following utility sectors: water, wastewater, solid waste, and certain types of energy efficiency projects. Those six chapters are written from the perspective of a utility executive who needs a project to improve his system and is interested in learning the financial techniques necessary to obtain the money to carry out the project.

After introducing the concept of project finance (Chapter 1), the first step was the presentation of methods for measuring income (Chapter 2 – Measuring Income). Methods for maximizing income were presented in the succeeding chapter, Maximizing Cash Available for Debt Service. It was then emphasized that the maximization of income was of paramount importance because if income exceeded operating expenses, needed projects could be undertaken to improve utility systems by having the utility incur debt. If they had excess income, they would be free of a pernicious reliance on grants. Excess income can be used to pay annual debt service. The next chapter, therefore, was devoted to the explanation of the elemental concepts of loans and debt (Chapter 4 – Loan Basics).

Once the basic concepts of borrowing are understood, the next step is to be able to evaluate various types of loans in order to determine which available loan is best for a particular utility. Techniques for evaluating loans, therefore, were presented in the following chapter, Project Valuation.

The next chapter used all information presented in the preceding five to illustrate which projects were financially feasible for a utility to undertake. It demonstrated both how the largest possible project could be undertaken at the lowest possible cost and how a utility's excess operating income could be used to improve the system (Chapter 6 – Financial Feasibility).

In Chapter 7 – Alternate Finance Sources the perspective shifts. Financial concepts were no longer presented from the utility executive's perspective. Rather, in this chapter, financial concepts were presented from the perspective of a government official whose responsibility it is to manage or administer a program to provide funds for municipal utility services. In other words, in the first six chapters, project finance was discussed from the perspective of one needing funds (for a particular project). In the seventh chapter, the concepts of project finance were discussed from the perspective of a government official who has funds, or whose job it is to seek funds for utility projects in his country. It is this official's responsibility to provide the best quality of utility services to the largest amount of people by providing funds to utility operators for projects which improve that utility's services to its users.

In the eighth chapter, Sources of Funds, the same perspective was maintained as in the previous chapter. It was written for a government official whose responsibility is to manage a program to finance projects for utilities. In that chapter, sources of external funding for such programs were examined and compared with a view to the most likely and most reliable sources of funding for utility projects.

In Chapter 9 – Cost/Benefit Analyses, government officials were presented with a basic rubric of how to optimize often meager funds available for environmental projects. Matrices were devised to prioritize the "projects which bring the greatest amount of public health benefits to the largest number of people for the least amount of grant funds."

In this chapter, Tariff Design, we return to addressing the needs of local government officials and utility managers. In Chapter 3 – Maximizing Cash Available for Debt Service, the need for creating fair, volumetric tariffs that provided enough income to cover the cost of operating and maintaining the utility system was discussed. In this chapter, the methodology of creating such tariffs is presented. This is also a discussion of how such tariffs can be regulated by local governments to assure users of the basic fairness of the tariff design.

CHAPTER CONTENTS

This chapter will discuss the following four topics as sources of funding for environmental utility and other utility type of funding:

- Decision Making Objectives
- Full Cost Recovery Tariffs
- Tariff Design Options
- Regulation of Water Tariffs

DECISION MAKING OBJECTIVES

Objectives of Municipal Tariff Design

Setting water tariffs requires a balance between four main objectives:

1. Cost Recovery
2. Economic Efficiency
3. Equity
4. Affordability

2

A utility manager is faced with difficult and often competing objectives when it comes to making decisions regarding the price to charge customers for water use (i.e. the water tariff), these include:

- *Revenue Sufficiency/Cost Recovery*: tariffs produce stable revenue that is equal to the financial cost of supplying the water service.
- *Economic Efficiency*: water prices signal to consumers the financial, environmental, and other costs that their decisions to use water impose on the rest of the system and on the economy.
- *Equity*: equals are treated equally – in other words, the prices charged to customers are equal to the costs imposed on the system by those customers.
- *Resource Conservation*: pricing decisions should not promote the unwise use of water resources.
- *Net Revenue Stability*: prices should allow the utility to have sufficient income to meet its operating costs, even when quantities demanded are below a normal level.
- *Transparency*: pricing structures should be able to be understood by every consumer, in order for the consumer to respond accurately when deciding how much water to consume.
- *Ease of Implementation*: the pricing structure should not impose significant administrative costs on the utility.

- *Affordability*: the prices charged to customers should be within a standard limit of affordability.

Objectives of Municipal Tariff Design

1. Cost Recovery

- From the water supplier's point of view, cost recovery is the main purpose of the tariff
- Cost recovery requires that, on aggregate, tariffs produce stable revenue equal to the financial cost of the supply

3

Objectives of Municipal Tariff Design

2. Economic Efficiency

- Requires that prices signal to consumers the financial, environmental, and other costs that their decisions to use water impose of the rest of the system and on the economy
- The volumetric charge should be set equal to the marginal cost of bringing one additional cubic meter of water into a city and delivering it to a particular customer

4

Objectives of Municipal Tariff Design

3. Equity

Equity means users pay monthly water bills that are proportionate to the costs they impose on the utility by their water use.

5

Objectives of Municipal Tariff Design

4. Affordability

- Many people feel that because water services have a major impact on health and well being they should be provided to people regardless of whether they can pay for them
- However, in practice, somebody must pay for water services, either the taxpayer or other customers
- Providing water free would conflict with the objectives of cost recovery and efficient water use

6

The objectives listed provide a guideline for utility managers to follow when deciding on water tariffs; however, not all of the objectives can be met at the same time. This chapter will focus on two of the objectives that are most directly related to the financial stability of the utility: cost recovery and economic efficiency. The objectives of equity and affordability will be discussed in the next chapter, Subsidies.

FULL COST RECOVERY TARIFFS

When setting the tariff, utility managers must understand that they must at least be able to recover the cost of operating the water system from the revenues earned. If the utility did not have to provide for the cost of building a new project (i.e. the project costs), then the tariff must be set at the level where it will equal the amount of total cash expenses. One method for designing a tariff is to begin with the operating budget (equal to the total cash expenses) and divide that amount by the total number of households that will purchase the water service; the result is the per household tariff required for the tariff to recoup the costs of operating the system.

It is highly unlikely that a utility can rely on grant funding for 100% of its project costs; it is probable that the utility will have to pay at least a portion of the project costs. The more project costs that are paid for by the utility, the higher the household water tariffs, since the utility must pass on the costs of providing services to the customers that use water. A helpful exercise will be to determine what the tariff must be for a community that is in need of a project to modernize its water system, so that the full cost of water delivery is recouped by the utility; this exercise is considered for different scenarios: 100% of the project funded by an outside grant; 75% of the project funded by a grant and 25% funded with a loan; 50% of the project funded by a grant and 50% by a loan; and 25% funded by a grant and 75% by a loan. Under these different scenarios, the important factors will be the term of the loan and the interest rate. Since income must equal the expenses (operating plus project costs), it is important to consider the effect of rate and term on the household tariff; it is possible to reduce expenses just as it is possible to increase income (i.e. the tariff) so that the tariff does not become unaffordable.

In the following example, assume that your utility serves 1,000 households and your current expenses[1] of $10,000 per year are covered completely by a $10 per household tariff. In order to modernize the water system and to repair pressing problems, a project is proposed that will cost $1,500 plus an increase in annual operating expenses of $550. Below are a few possible effects on the tariff corresponding to different project financing options.

Financing Option	Annual Operating Cost ($)	Annual Project Cost[2] ($)	Total Annual Cost ($)	Full Cost Recovery Tariff ($ per HH)	Percent Change in Tariff
100% Grant	*10,550*	*0*	*10,550*	*10.55*	*5.5*
75% Grant, 25% Loan (five year, 10%)	*10,550*	*112.5*	*10,662.50*	*10.66*	*6.63*
75% Grant, 25% Loan (ten year, 5%)	*10,550*	*56.25*	*10,606.25*	*10.61*	*6.06*
50% Grant, 50% Loan (five year, 10%)	*10,550*	*225*	*10,775*	*10.78*	*7.75*
50% Grant, 50% Loan (ten year, 5%)	*10,550*	*112.50*	*10,662.50*	*10.66*	*6.63*
25% Grant, 75% Loan (five year, 10%)	*10,550*	*337.50*	*10,887.50*	*10.89*	*8.88*
25% Grant, 75% Loan (ten year, 5%)	*10,550*	*168.75*	*10,718.75*	*10.72*	*7.19*
100% Loan (five year, 10%)	*10,550*	*450*	*11,000*	*11.00*	*10.0*
100% Loan (ten year, 5%)	*10,550*	*225*	*10,775*	*10.78*	*7.75*

As expected, the more that a utility must pay for the project costs in a full cost recovery tariff system, the larger the impact on the household tariff. As shown above, a longer-term loan at a much lower interest rate has a smaller impact on the costs (and therefore tariff) than does a shorter term loan.

1 Expenses include: energy, labor, chemicals, and administration costs.
2 Using a level principle payment schedule, the total annual payment decreases annually; in this example, the annual project cost is equal to the first year's payment, therefore subsequent year payments will be lower.

TARIFF DESIGN OPTIONS

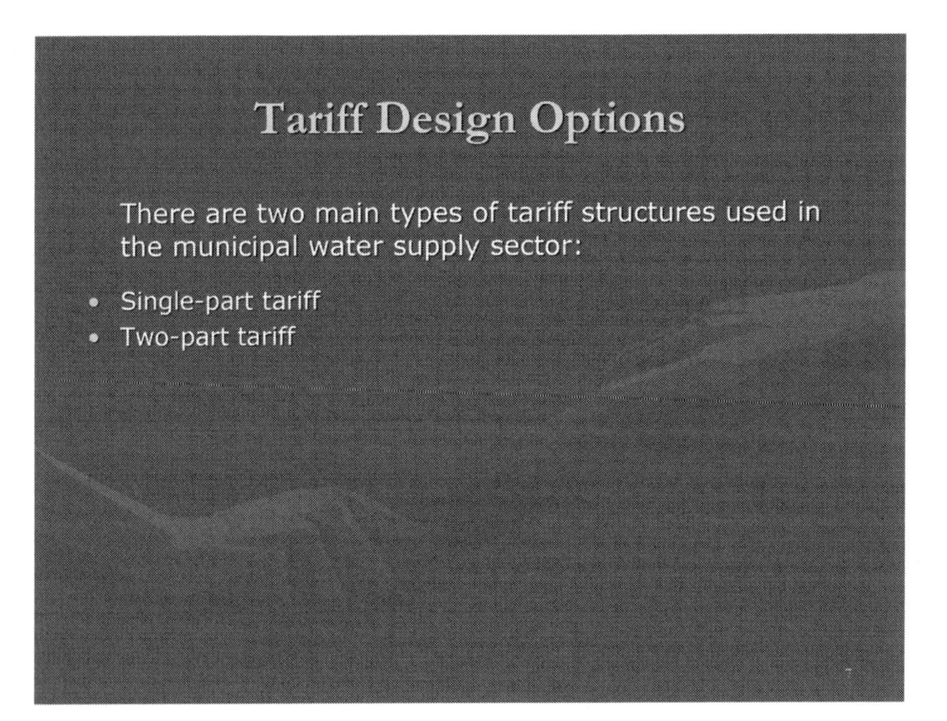

Given the requirement that utilities use full cost recovery methods as described above, there are a variety of ways that a utility can design the tariff scheme. To demonstrate the unique features of each tariff design option, the following example will be applied to each option:

A public utility has 100 customers that purchase the following amounts of water each month:

25 customers consume 500 cubic meters of water
50 customers consume 1000 cubic meters of water
25 customers consume 1500 cubic meters of water

Fixed Vs. Volumetric Charges

Tariff Design Options

Single-Part Tariffs

- **Fixed charge:** monthly water bill is independent of the volume consumed
- **Water use charge**
  - Uniform volumetric tariff
  - Block tariff: unit charge is constant over a specified range of water use and then shifts as use increases
    - Increasing Block
    - Decreasing Block
  - Increasing linear tariff: unit charge increases linearly as water use increases

8

The first way that we will distinguish tariffs is by those that have a fixed charge and those that have a volumetric charge.

> *Fixed Charge*: the amount of money that is charged to a customer is the same each month, and is independent of the amount of water used by each customer.

In places where there are no water meters, volumetric charges are not a viable option – because there is no way to accurately measure how much water each customer consumes. In this situation, a *single-part tariff* using fixed charges is the only available option.

> *Single Part Tariff:* a tariff structure that has only one method for assigning charges, either a fixed charge used alone or a volumetric charge used alone.

When setting the fixed charge, or fee, for all customers, the utility must first determine its own costs for delivering the amount of water demanded by all customers, because the total revenue that the utility will receive must cover the utility's total costs plus provide funds for maintenance and repairs. A utility may define different "classes" of customers, so that residential customers are charged different (usually less) fixed rates than business customers since businesses tend to consume more water than residential users.

Fixed charges, when used alone, have two major drawbacks. Firstly, consumers are given no incentive to economize on water use since each additional cubic meter of water comes free of charge. Secondly, as water use increases, as will happen with a growing population and a developing economy, the utility's ability to recover its costs by the fixed charge will diminish as the costs of meeting growing needs increases.

In our example, all 100 customers are charged the same fee (in this case it is set at $20 each month), even though they consume different amounts of water.

Consumption Level	Total Charge
500 cm	*$20*
1000 cm	*$20*
1500 cm	*$20*

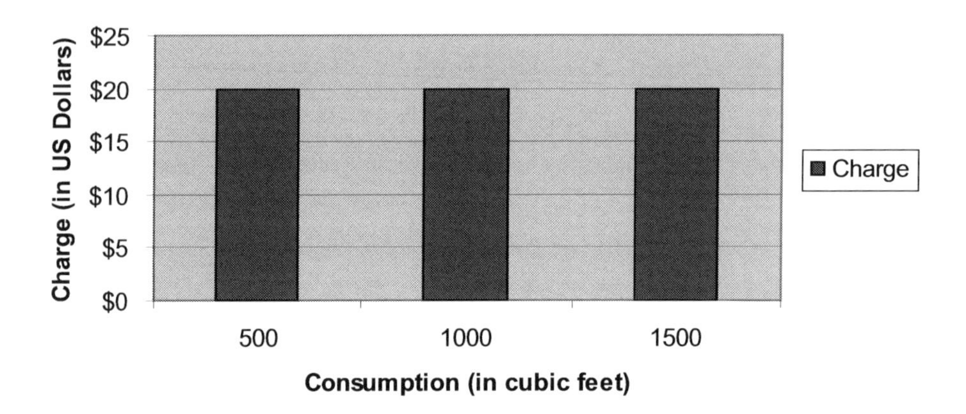

Volumetric Charge: the amount of money that is charged to each customer is dependent on the amount of water used.

A utility that has water meters on all or most household connections is able to monitor the amount of water that each consumer (i.e. household) uses and is thus able to charge customers according to the amount of water consumed each month. When setting this price, the utility should attempt to set the volumetric charge at the *marginal cost* per unit of water provided to each consumer.

Marginal Cost: the cost incurred by the utility to provide an *additional* unit (i.e. cubic foot) of water supply to a customer.

The marginal cost estimate must incorporate administrative costs as well as the costs associated with each additional volume of water provided because in a single-part tariff, there is no additional fixed charge added to consumers' bills to cover the administrative costs.

To calculate the total charge for each customer group (in this case groups are separated by the amount of water consumed) multiply the amount of water consumed by the price per cubic feet of water.

Uniform Volumetric Charge: a volumetric tariff structure where the price per unit of water is the same regardless of how much water is purchased in total by each customer.

Tariff Design Options

Uniform Volumetric Charge

- With the uniform volumetric charge, the water bill is simply the quantity used (e.g. cubic meters) times price per unit of water (local currency per cubic meter)
- A uniform volumetric charge has the advantage that it is easy for consumers to understand
- It can be used to send a clear, unambiguous signal about the marginal cost of using water

14

In this example a *uniform volumetric charge* is used where each consumer group is charged the same volumetric price; the next section discusses other forms of volumetric charges.

$$\textbf{Total Charge} = \textbf{Amount Consumed (cubic feet)} * \textbf{Price per cubic foot}$$

The volumetric charge example is different from the fixed charge example in that the charges now depend on the amount of water consumed, so each customer group has a different total bill for water. The total charges are calculated as follows using a uniform volumetric charge of $0.02 per cubic meter:

Consumption Level	Total Charge Calculation	Total Charge
500 cm	*500 * $0.02*	*$10*
1000 cm	*1000 * $0.02*	*$20*
1500 cm	*1500 * $0.02*	*$30*

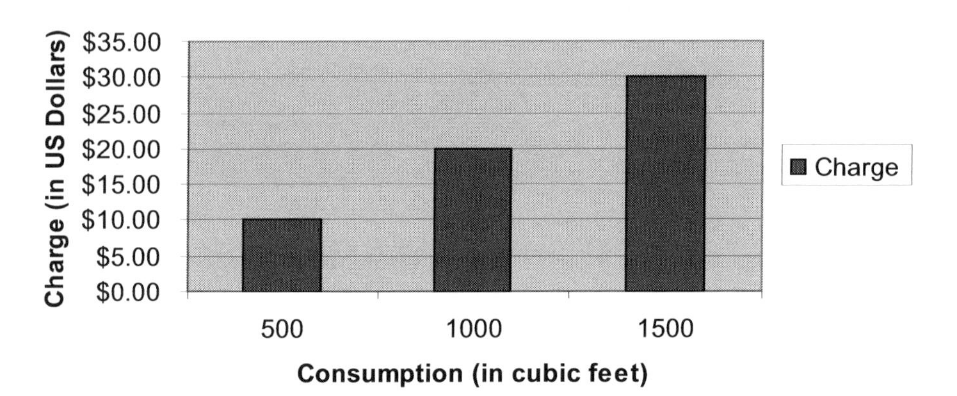

Multipart Tariff

Tariff Design Options

Two-Part Tariffs

Fixed Charge + Water Use Charge

9

Tariff Design Options

Two-Part Tariffs

With a two-part tariff, the consumer's water bill is based on the sum of two calculations:

1. Fixed charge
2. Charge related to the amount of water used

27

Tariff Design Options

Two-Part Tariffs (cont.)

There are many variations in the way these two components can be put together:

- The fixed charge can be either positive or negative (i.e., rebate)
- The water use charge can be based on any of the volumetric tariff structures previously described (i.e., a uniform volumetric tariff; an increasing or decreasing block tariff)

28

Tariff Design Options

Two-Part Tariffs (cont.)

- In many cases, the fixed charge is kept uniform across customers and relatively low in value, and is used simply as a device for recovering the fixed administrative costs associated with meter reading and billing which are unrelated to the level of water consumption
- Two-part tariffs enable water utilities to simultaneously achieve economic efficiency and cost recovery objectives

29

Tariff Design Options

- If a large capacity expansion project has recently been completed, the short-run marginal cost of raw water supply may be very low
- Economic efficiency requires that water be priced as a short-run marginal cost
- If this leads to a very low water price, it is likely that a single-part tariff will not recover the total cost of supply
- If a two-part tariff is used, the necessary revenues can be raised with a fixed charge

30

Tariff Design Options

- In periods of water scarcity, pricing at short-term marginal cost implies that the volumetric charge must include the opportunity cost to the user who does not receive water due to scarcity
- Scarcity causes volumetric charge to be rather high, which produces revenues in excess of financial costs
- This can be corrected by employing a negative fixed charge, providing customers with a rebate while the volumetric charge remains high enough to signal economic efficiency

31

A tariff can be made up of either a fixed or a volumetric charge, as explained above – or it can have both, which would be a multi-part tariff. In the case of a **multipart tariff**, the fixed charge element of the water bill tends to cover infrastructural costs (e.g. the cost of installing and reading water meters and billing customers) incurred by the utility in order for it to be able to deliver water to each customer and the volumetric charge amount reflects the "per unit" cost of delivering water to the consumer.

Multipart Tariff: a tariff structure that uses both a fixed and a volumetric charge.

The fixed charge is in addition to the price of delivering the amount of water demanded by that customer (i.e. the volumetric charge) because it reflects the costs incurred by the water utility to provide services to customers regardless of the level of water consumption.

Tariff Design Options

Fixed Charges

- In the absence of metering, fixed charges are the only possible tariff structure
- With the fixed charge, the consumers monthly water bill is the same regardless of the volume used
- It is common for businesses to have a fixed charge different from households, based on the assumption that:
  - Firms use more water than households, and
  - Firms have a higher ability to pay than households

10

Tariff Design Options

Fixed Charges (cont.)

- From an economic efficiency perspective, the problem with a fixed-charge system is that consumers have absolutely no incentive to economize on water use since each additional cubic meter comes free of charge
- A fixed charge that provides sufficient revenues at one point in time will become increasingly inadequate as the economy and incomes grow and water use increases

11

In our example, the fixed charge will be $10 per connection (i.e. household) and the uniform volumetric charge will be $0.02 per cubic foot of water consumed.

Consumption Level	Total Charge Calculation	Total Charge
500 cm	*$10 + (500 * $0.02)*	*$20*
1000 cm	*$10 + (1000 * $0.02)*	*$30*
1500 cm	*$10 + (1500 * $0.02)*	*$40*

Notice how the total charges are all simply $10 higher than the respective charges for the single part uniform volumetric charge.

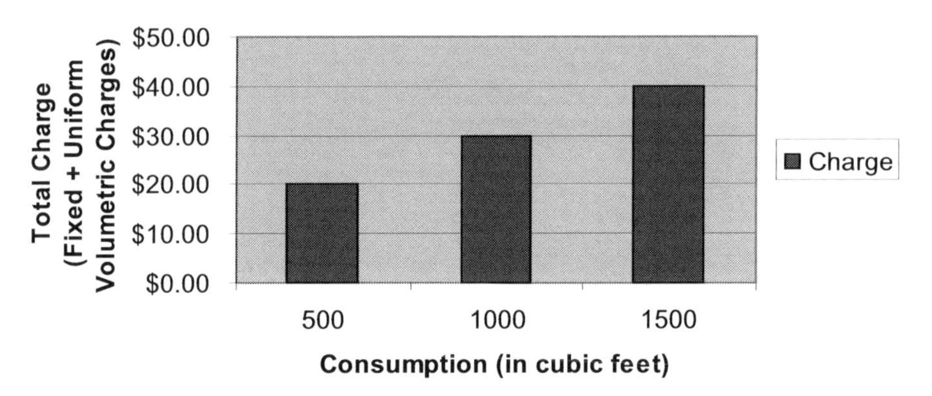

Types of Volumetric Charges

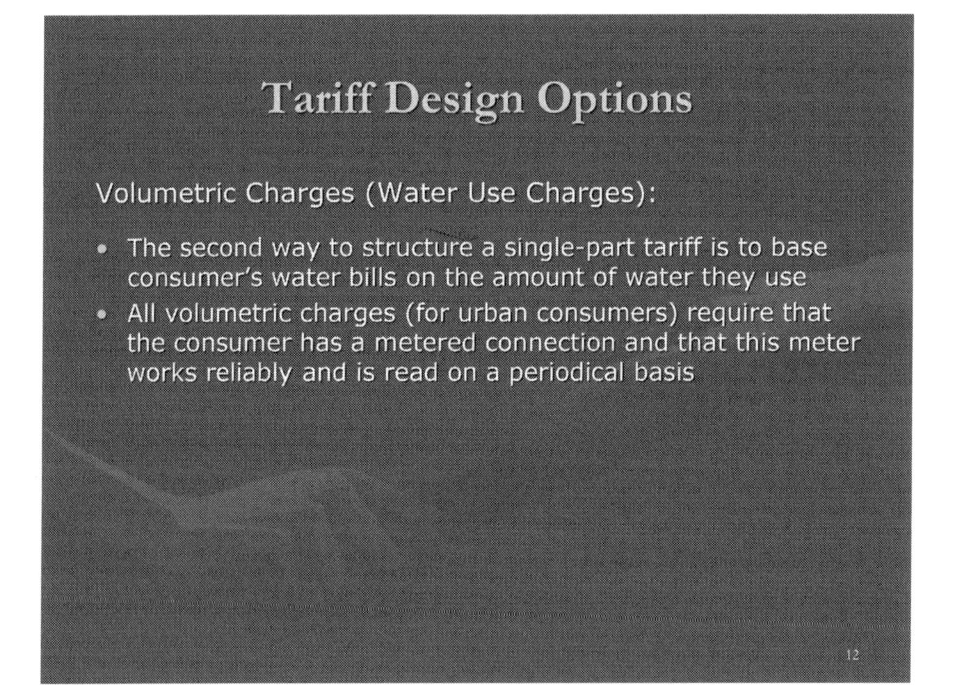

Tariff Design Options

Volumetric Charges (cont.)

There are three main options to calculate the water bill using the volumetric charge:

1. Uniform volumetric charge
2. Block tariff where the unit charge is specified over a range of water use for a specific customer, and then shifts as use increases
3. Increasing linear tariff where the unit charge increases linearly as water use increases

13

There are a small number of accepted methods for designing volumetric charges that must be considered when setting tariffs; so that the utility is able to select a tariff that will achieve full recovery of the costs to deliver water service, while trying to achieve the other goals of equity, affordability, and economic efficiency. The four volumetric pricing design options, in addition to the uniform volumetric charge that was described previously, are: increasing block tariff, declining block tariff, seasonal pricing, and zonal pricing.

Tariff Design Options

Block Tariffs come in two main varieties:

1. Increasing block tariff (IBT)
2. Decreasing block tariff (DBT)

15

Increasing Block Tariff: consumers face a low per-unit charge up to a specific quantity (or block), then any water consumed beyond that quantity will be charged at a higher price up to the limit of the second block, and so on for as many blocks as the tariff utilizes.

Tariff Design Options

Block Tariffs – IBT

- With IBT, consumers face a low volumetric per-unit charge (price) up to a specified quantity (or block); and then for any water consumed in addition to this amount they pay a higher price up to the limit of the second block, and so on
- IBT's are widely used in countries where water resources have historically been scarce

16

The increasing block tariff (IBT) is commonly used in countries where water resources have been historically scarce. In theory, IBT can achieve three objectives simultaneously:

1. Promote affordability by providing the poor with affordable access to a "subsistence block" of water;
2. Achieve efficiency by confronting consumers in the highest price block with the marginal cost of using water;
3. Raise sufficient revenues to recover costs.

Tariff Design Options

In theory, IBT can achieve three objectives simultaneously:

1. Promote affordability by providing the poor with affordable access to a "subsistence block" of water
2. Achieve efficiency by confronting consumers in the highest price block with the marginal cost of using water
3. Raise sufficient revenues to recover costs

24

In practice, however, IBTs often fail to meet any of the three objectives mentioned above, in part because they tend to be poorly designed. Many IBTs fail to reach cost recovery and economic efficiency objectives, usually because the upper consumption blocks are not priced at sufficiently high levels and/or because the first subsidized block is so large that almost all residential consumers never consume beyond this level.

Tariff Design Options

- In practice, IBTs often fail to meet any of the three objectives previously mentioned, in part because they tend to be poorly designed
- Many IBTs fail to reach cost recovery and economic efficiency objectives, usually because the upper consumption blocks are not priced at sufficiently high levels and/or because the first subsidized block is so large that almost all residential consumers never consume beyond this level

25

When used in a multipart tariff, the increasing block pricing scheme affects only the volumetric part of the total tariff; the fixed charge remains the same over all customer groups. Calculating the total charge for each customer group using an increasing (or a declining) block tariff first requires delineating the number of blocks and the quantity of water allowed in each block. Customers will pay the fixed charge plus the sum of the products of the amount of water consumed in each block multiplied by the per unit price of water in each block.

Total Charge = Fixed Charge

+ [(Amt. Consumed in Block 1 * Unit Price for Block 1)

+ (Amt. Consumed in Block 2 * Unit Price for Block 2)

+ (Amt. Consumed in Block 3 * Unit Price for Block 3)]

In our example, the three blocks will be as follows:

Block 1: $0.01 per cubic meter for first 500 cm of water consumed

Block 2: $0.02 per cubic meter for amounts between 501 cm and 1000 cm of water consumed
Block 3: $0.03 per cubic meter for amounts over 1000 cm of water consumed

Consumption Level	Total Charge Calculation	Total Charge
500 cm	*$10 + (500 * $0.01)*	*$15*
1000 cm	*$10 + [(500 * $0.01) + (500 * $0.02)]*	*$25*
1500 cm	*$10 + [(500* $0.01) + (500* $0.02) + (500* $0.03)]*	*$40*

Notice how the incremental increase of 500 cm of water resulted in an exponentially higher total charge for the third block than for the second block.

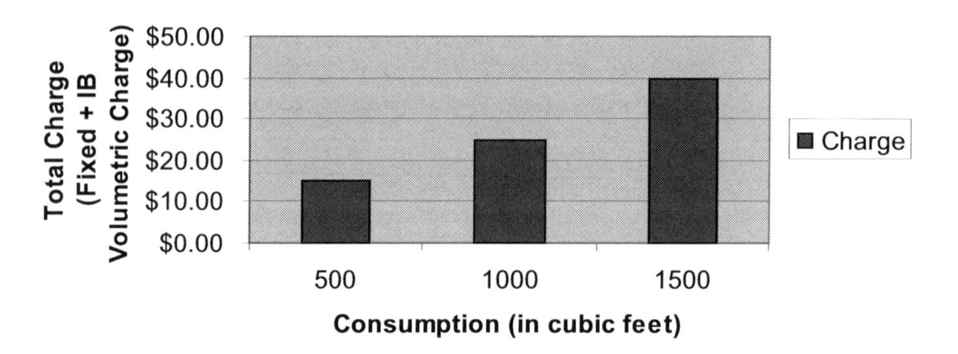

The next tariff design option to be discussed is the declining block tariff (DBT).

Declining Block Tariff: the opposite of the increasing block tariff; consumers face a high volumetric charge up to the specified quantity of the first block, then any water consumed beyond this level (and up to the next block) is charged at a lower rate, and so on for as many blocks as the tariff utilizes.

Tariff Design Options

Block Tariffs – DBT

With DBT, on the other hand, consumers face a high volumetric charge up to the specified quantity in the first block, and then for any water consumed in addition to this amount, they pay a lower price up to the limit of the second block, and so on.

17

The DBT structure was designed to reflect the fact that when raw water supplies are abundant, large industrial customers often impose lower average costs because they enable the utility to capture economies of scale in water source development, transmission and treatment. This tariff design has gradually fallen out of favor, in part because marginal costs, properly defined, are now relatively high in many parts of the world, and there is thus increased interest in promoting water conservation by the largest customers. The declining block tariff structure is also often politically unattractive because it results in high volume users paying lower than average water prices.

Tariff Design Options

- The DBT structure was designed to reflect the fact that when raw water supplies are abundant, large industrial customers often impose lower average costs because they enable the utility to capture economies of scale in water source development , transmission, and treatment
- The DBT has gradually fallen out of favor, in part because marginal costs, properly defined, are now relatively high in many parts of the world, and there is thus increased interest in promoting water conservation by the largest customers
- The DBT structure is also often politically unattractive because it results in high volume users paying lower average water prices

26

In the declining block rate example, we will continue to use a fixed rate of $10 and the block delineations; however, the per unit price of water is now:

Block 1: $0.03 per cubic meter for first 500 cm of water consumed
Block 2: $0.02 per cubic meter for amounts between 501 cm and 1000 cm of water consumed
Block 3: $0.01 per cubic meter for amounts over 1000 cm of water consumed

Consumption Level	Total Charge Calculation	Total Charge
500 cm	*$10 + (500 * $0.03)*	*$25*
1000 cm	*$10 + [(500 * $0.03) + (500 * $0.02)]*	*$35*
1500 cm	*$10 + [(500* $0.03) + (500* $0.02) + (500* $0.01)]*	*$40*

Notice how the total charge for the first two groups' users is now $10 higher than with the increasing block tariff, whereas the total charge for the largest water user did not change.

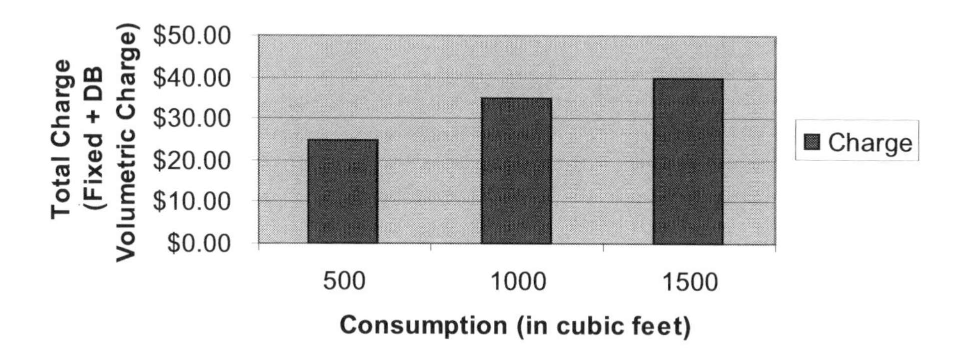

Tariff Design Options

For both IBT and DBT the water bill is calculated in the following manner:

- Let Q^* = amount of water sold to a specific customer
- Q_1 = maximum amount of water that can be sold to a consumer in the first block at P_1
- Q_2 = maximum amount of water that can be sold to a consumer in the second block at P_2
- Q_3 = maximum amount of water that can be sold to a consumer in the third block at P_3

18

Tariff Design Options

Calculating IBT and DBT:

- If $Q^* < Q_1$, then the consumer's water bill = $(Q^*)P_1$
- If $Q_1 < Q^* < Q_2$, then the consumer's water bill = $P_1Q_1 + (Q^* - Q_1)P_2$
- If $Q_1 + Q_2 < Q^* < Q_3$, then the consumer's water bill = $P_1Q_1 + P_2Q_2 + (Q^* - [Q_1 + Q_2])P_3$
- And so on, for however many blocks there are in the tariff structure

19

Tariff Design Options

Example: Calculating **IBT**

- $Q_1 = 10m^3$ at $P_1 = 2\$/m^3$
- $Q_2 = 15m^3$ at $P_2 = 3\$/m^3$
- $Q_3 = 30m^3$ at $P_3 = 5\$/m^3$

20

Tariff Design Options

Example: Calculating **IBT** (cont.)

1. $Q^* = 9m^3$, find P^*
2. $Q^{**} = 13m^3$, find P^*
3. $Q^{***} = 26m^3$, find P^{***}

$$P^* = (Q^*)P_1 = (9m^3)2\$/m^3 = 18\$$$

$$P^{**} = P_1Q_1 + (Q^{**} - Q_1)P_2 = (10m^3)2\$/m^3 + (13m^3 - 10m^3)3\$/m^3 = 20\$ + 9\$ = 29\$$$

$$P^{***} = P_1Q_1 + P_2Q_2 + (Q^{***} - [Q_1 + Q_2])P_3 = (10m^3)2\$/m^3 + (15m^3)3\$/m^3 + (26m^3 - [25m^3])5\$/m^3 = 20\$ + 45\$ + 5\$ = 70\$$$

21

Tariff Design Options

Example: Calculating **DBT**

- $Q_1 = 10m^3$ at $P_3 = 5\$/m^3$
- $Q_2 = 15m^3$ at $P_2 = 3\$/m^3$
- $Q_3 = 30m^3$ at $P_1 = 2\$/m^3$

22

Tariff Design Options

Example: Calculating **DBT** (cont.)

1. $Q^* = 9m^3$, find P^*
2. $Q^{**} = 13m^3$, find P^*
3. $Q^{***} = 26m^3$, find P^{***}

$$P^* = (Q^*)P_1 = (9m^3)5\$/m^3 = 45\$$$

$$P^{**} = P_1Q_1 + (Q^{**} - Q_1)P_2 = (10m^3)5\$/m^3 + (13m^3 - 10m^3)3\$/m^3 = 50\$ + 9\$ = 59\$$$

$$P^{***} = P_1Q_1 + P_2Q_2 + (Q^{***} - [Q_1 + Q_2])P_3 = (10m^3)5\$/m^3 + (15m^3)3\$/m^3 + (26m^3 - [25m^3])2\$/m^3 = 50\$ + 45\$ + 2\$ = 97\$$$

23

In some circumstances the marginal cost of supplying water to customers may vary by seasons. In such cases, water tariffs can be used to signal customers that the costs of water supply are not constant across the seasons. Summer water use tends to be much higher for households that have gardens and any other high water uses; the increase in water use is usually found to be in outdoor water use, while indoor water use tends to remain constant throughout the year.

> *Seasonal Pricing*: when the marginal cost to provide water services changes according to seasons, then utilities can charge higher prices during the more costly (usually summer) and lower prices during the less expensive seasons (usually winter).

Tariff Design Options

Seasonal and Zonal Water Pricing

- In some circumstances, the marginal cost of supplying water to customers may vary by seasons
- In such cases, water tariffs can be used to signal customers that the costs of water supply are not constant across the seasons

32

Tariff Design Options

Seasonal and Zonal Water Pricing (cont.)

- Similarly, it may cost the water utility more to deliver water to outlying communities due, for example, to higher elevations and increased pumping costs
- Zonal prices can be used to ensure that users receive the economic signal that living in such areas involves substantially higher water supply costs
- However, this type of special tariff is only appropriate if the costs to serve the area are significantly higher than for the rest of the community – in fact costs vary among all users, and a practical tariff always reflects average costs to some degree

33

In this example, we will return to using the uniform volumetric tariff that charges the same per unit amount for water regardless of the amount of water consumed so that we can focus on the effect of a higher price for the summer season than for the winter season. We will assume that the amount of water consumed at all levels is constant during the year. The fixed charge is $10 and the seasonal prices are as follows:

Winter: $0.01 per cubic meter of water consumed
Summer: $0.02 per cubic meter of water consumed

Consumption Level – Summer	Total Charge Calculation	Total Charge
500 cm	*$10 + (500 * $0.02)*	*$20*
1000 cm	*$10 + (1000 * $0.02)*	*$30*
1500 cm	*$10 + (1500 * $0.02)*	*$40*

Consumption Level – Winter	Total Charge Calculation	Total Charge
500 cm	*$10 + (500 * $0.01)*	*$15*
1000 cm	*$10 + (1000 * $0.01)*	*$20*
1500 cm	*$10 + (1500* $0.01)*	*$25*

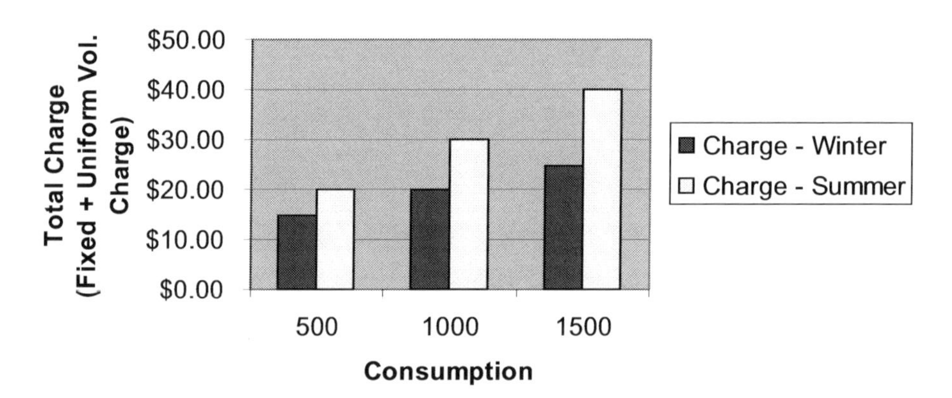

A seasonal and/or zonal pricing scheme can be added to any of the three pricing methods above (uniform, IBT and DBT) as they are simply adjustments to the chosen pricing scheme based on the time of the year (seasonal pricing) and the location of customers (zonal pricing).

Zonal Pricing: when the marginal cost to provide water services varies according to the location of the customer, utilities can charge higher volumetric charges to the customers that live in an area that is more costly to serve, and lower volumetric charges to those that live in an area that is cheaper to serve.

It may cost the water utility more to deliver water to outlying communities due, for example, to higher elevations and increased pumping costs. Zonal prices can be used to ensure that users receive the economic signal that living in such areas involves substantially higher water supply costs. However, this type of special tariff is only appropriate if the costs to serve the area are significantly higher than for the rest of the community – in fact costs vary among all users, and a practical tariff always reflects average costs to some degree.

In the zonal pricing example, we will continue to use a fixed charge of $10 and a uniform volumetric charge with variations in the per unit charge depending on the zone in which consumers reside.

Zone 1: $0.02 per cubic meter of water consumed
Zone 2: $0.03 per cubic meter of water consumed

Consumption Level – Zone 1	**Total Charge Calculation**	**Total Charge**
500 cm	*$10 + (500 * $0.02)*	*$20*
1000 cm	*$10 + (1000 * $0.02)*	*$30*
1500 cm	*$10 + (1500 * $0.02)*	*$40*

Consumption Level – Zone 2	**Total Charge Calculation**	**Total Charge**
500 cm	*$10 + (500 * $0.03)*	*$25*
1000 cm	*$10 + (1000 * $0.03)*	*$40*
1500 cm	*$10 + (1500 * $0.03)*	*$55*

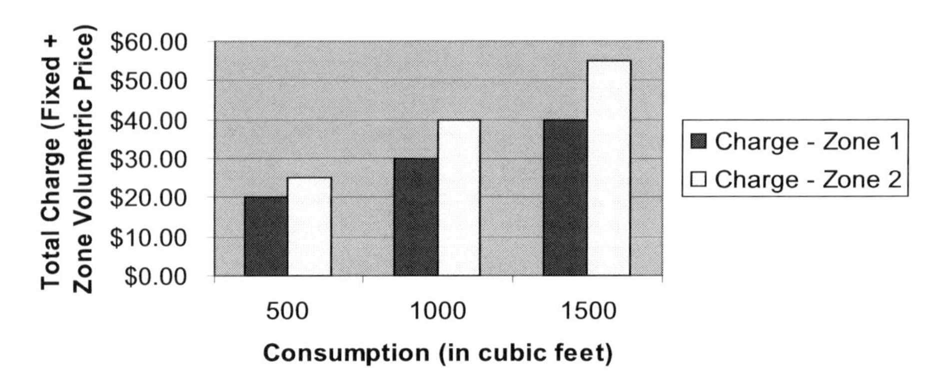

The goals of full cost recovery and economic efficiency are achieved when utilities charge customers according to the cost that each customer imposes on the system. If the cost to deliver water is the same for all of the utility's customers, then a uniform volumetric charge that is set equal to the cost of water delivery is the most efficient. The DBT is used when the cost to deliver large amounts of water is cheaper than for small amounts of water. When the marginal cost of water is high, IBT is usually chosen over a decreasing block tariff since it charges higher rates to customers that demand higher levels of water.

REGULATION OF WATER TARIFFS

There are many different types of utility systems throughout the world including public agencies (local or national), government corporations, private but regulated companies, and private and unregulated companies. These tend to be larger utility systems, while there are also small systems (usually rural) that are governed by local water committees. The size and legal status (i.e. public or private) of the utility system are important indicators of whether or not the utility system will need public regulation of the prices charged to customers.

- Community (Small) Water Systems: are self regulating, since the community comes together to determine how much money to charge each household for access to the community system.
- Medium and Large Urban Water Systems: may require price regulations to protect the public from unfair pricing.

Public (i.e. governmental) water utilities set prices at levels that benefits society the most; this often means that prices are set so that the utility can recoup the costs of providing the service plus a small amount to cover future repairs and maintenance; another requirement for public utilities is that all or almost all of the local population is covered by the water service. Prices of water that is delivered by private water companies in the absence of price regulation tend to be much higher because such companies are often *natural monopolies*.

Natural Monopoly: when one firm can produce all demand of a product or service at a lower cost than any other firm; is true for water utilities since the cost for a new firm to enter the market is so high that no firms will attempt to compete with the monopolist, who likely owns or has unrestricted access to the entire water system.

The goal of a natural monopoly is often to maximize its own profits, which is easily accomplished because of its market power. Therefore, the monopoly will charge the highest price possible for any amount of water it delivers. This price is likely going to be too high for a significant part of the market to afford, which leaves the poorest people without this necessary service. Enforceable regulation of private water companies can correct these market failures.

Types of Regulation

There are two general types of regulation: economic and social. Economic regulation generally includes any policy or rule that governs the prices set by private companies, and the types and amount of goods or services supplied by those companies. Social regulation includes all other factors that are not covered by economic regulation; with respect to water utilities, this would include water quality standards. Both types require that an independent governmental agency be responsible for monitoring the prices charged by the utility and the quality of the water delivered by the utility; it is not expected that one agency should take on both roles although this might be necessary in a small water system. The important aspect of regulation that is often lacking in many developing countries is the ability of regulating agencies to enforce their decisions. Enforcement can only occur where there is a legal basis for enforcement and sanctions for noncompliance must be of enough value to prevent the utility from engaging in illegal or inappropriate activities.

The choice of how to regulate water prices must be made by the government with input from the utility, consumer advocates, and other interested parties. Regulating agencies have many available methods of deciding the appropriate price for the utilities to charge customers; however, the important rule is that utilities must be able to recoup their necessary costs of providing the service and be allowed to retain a small amount above their costs in order to save for future repairs and maintenance. Annual adjustments should be allowed, such as an annual increase equal to the rate of inflation, without requiring any extensive appeal to the regulatory agency.

Independent Financial and Management Audit

Monitoring of utility pricing and water quality is the most important role of the regulating agency. The utilities should not be allowed to report their prices and water quality measures to the regulators without an independent monitor to verify the accuracy of the data; this is the only way to ensure that society's interests are protected. The regulatory agency and the utility should agree upon a third part – a truly independent group of monitors – to conduct the annual economic and water quality surveys of the utility. A third party is most likely to provide an objective analysis that will benefit both the regulator and the utility.

Tariff Design Options

Table 1: Water Tariff Structures (as share of utilities)

Country	Fixed Charge	Uniform Volumetric Charge	Increasing Block Tariff	Decreasing Block Tariff
Australia	-	68%	27%	5%
Canada	56%	27%	4%	13%
France	2%	98%	-	-
Hungary	-	95%	5%	-
Japan	-	42%	57%	1%
Turkey	-	-	100%	-
UK	90%	10%	-	-
US	2%	33%	31%	34%
Sweden	-	100%	-	-

34

Tariff Design Options

It is clear (from Table 1) that there is wide variation on tariff setting practices around the world, and that there is no consensus on which tariff structure best balances the objectives of the utility, consumers, and society.

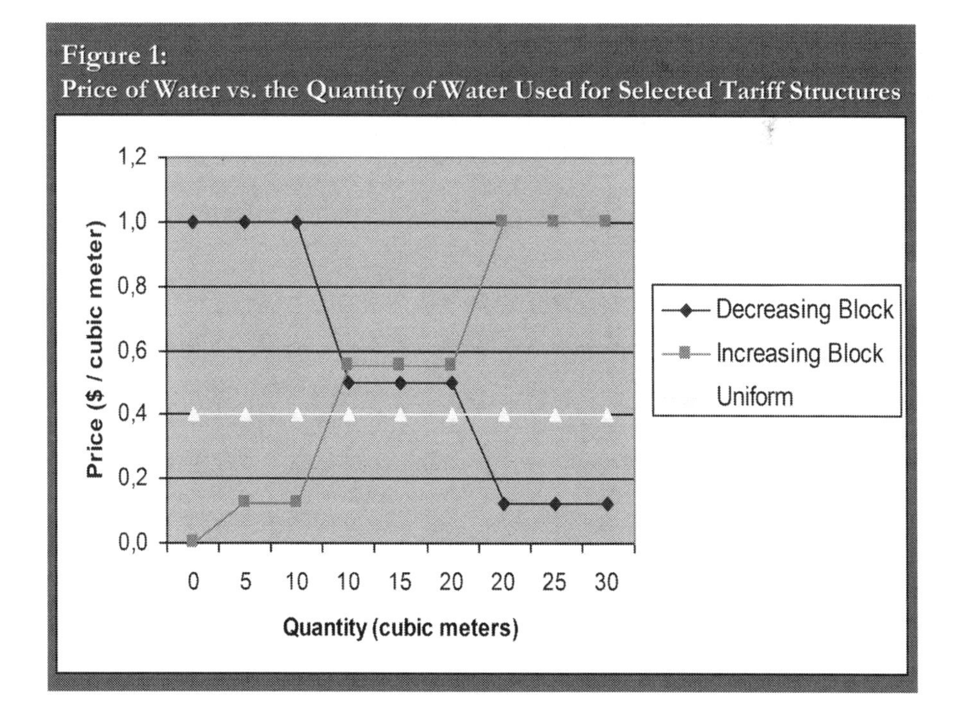

Figure 1:
Price of Water vs. the Quantity of Water Used for Selected Tariff Structures

Table 2:
Summary of Alternative Tariff Structures Against Design Objectives

Tariff Structure	Objectives Cost Recovery	Economic Efficiency	Equity	Affordability
Fixed charge	**Adequate** Provides stable cash flow if set at appropriate level, but utility may be vulnerable to resale of water and spiraling consumption.	**Poor** Does not send a message about the cost of the additional water.	**Poor** People who use large quantities of water pay the same as those who use little.	**Adequate** If differentiated by ability to pay, but households are unable to reduce their bills by economizing on water use.

Tariff Structure	Objectives Cost Recovery	Economic Efficiency	Equity	Affordability
Uniform Volumetric Charge	**Good** If set at appropriate level, moreover revenues adjust automatically to changing consumption.	**Good** If set at or near marginal cost of water.	**Good** People pay according to how much they actually use.	**Good** Can be differentiated by ability to pay, and people can limit their bills by reducing consumption.

	Objectives			
Tariff Structure	**Cost Recovery**	**Economic Efficiency**	**Equity**	**Affordability**
Increasing Block Tariff	**Good** But only if the size and height of the blocks are well designed.	**Poor** Typically little water is actually sold at marginal cost.	**Poor** People do not pay according to the costs their water use imposes on the utility.	**Poor** Penalizes poor families with large households and/or shared connections.

	Objectives			
Tariff Structure	**Cost Recovery**	**Economic Efficiency**	**Equity**	**Affordability**
Decreasing Block Tariff	**Good** But only if the size and height of the blocks are well designed.	**Poor** Typically little water is actually sold at marginal cost.	**Poor** People do not pay according to the costs their water use imposes on the utility.	**Poor** Penalizes poor families with low levels of consumption.

Tariff Design Options

In most cases, the performance of each type of tariff structure against the four key performance objectives previously discussed depends not only on the choice of tariff structure but on the level at which the tariff is set, and whether or not some kind of subsidy scheme is built in to address the issue of affordability.

41

Questions and Answers

Question #1:

As a utility manager, when it comes to making decisions regarding the price to charge customers for water use, which of the following objectives would not be used as a guideline:

A) Affordability
B) Cost Recovery
C) Economic Efficiency
D) Marginal Cost
E) Equity

Question #2:

Assume that your utility serves 8,000 households and your current expenses of $120,000 per year are covered completely by a $15 per household tariff. In order to modernize the water system and make essential repairs, a project is proposed that will cost $10,000, plus an increase in annual operating expenses of $2500. Calculate the possible effects on the tariff corresponding to the different project financing options and populate your answers within the below table.

Note: When calculating the Annual Project Cost, using a level principal payment method (refer to Chapter 5), the total annual payment decreases annually; in this example, the annual project cost is equal to the first year's payment, therefore subsequent year payments will be lower.

Financing Option	Annual Operating Cost ($)	Annual Project Cost ($)	Total Annual Cost ($)	Full Cost Recovery Tariff ($ per HH)	Percent Change in Tariff
100% Grant					
75% Grant, 25% Loan (seven year, 15%)					
75% Grant, 25% Loan (14 year, 8%)					
50% Grant, 50% Loan (seven year, 15%)					
50% Grant, 50% Loan (14 year, 8%)					
25% Grant, 75% Loan (seven year, 15%)					
25% Grant, 75% Loan (14 year, 8%)					
100% Loan (seven year, 15%)					
100% Loan (14 year, 8%)					

Question #3:

A public utility has 175 customers that purchased the following amount of water for the month of June:

- Group A: 40 customers consume 600 cubic meters of water
- Group B: 60 customers consume 900 cubic meters of water
- Group C: 75 customers consume 1400 cubic meters of water

Note: Use the information above to answer all parts of Question #3.

Part A:

Assuming all 175 customers are charged a fixed tariff of $30 each month for water consumed during the month of June, what is the total charge for each group?

Part B:

Assuming all 175 customers are charged a uniform volumetric tariff of $0.04 per cubic meter for water consumed during the month of June, what is the total charge for each group?

Part C:

Assuming all 175 customers are charged a multipart tariff of $15 per connection (i.e. household) and a uniform volumetric charge of $0.04 per cubic meter for water consumed during the month of June, what is the total charge for each group?

Part D:

Assuming all 175 customers are charged a multipart tariff of $15 per connection and an increasing block volumetric charge for water consumed during the month of June, what is the total charge for each group?

The three blocks are as follows:

- Block 1: $0.02 per cubic meter for first 600 cm of water consumed
- Block 2: $0.04 per cubic meter for amounts between 601 cm and 1100 cm of water consumed
- Block 3: $0.06 per cubic meter for amounts over 1100 cm of water consumed

Part E:

Assuming all 175 customers are charged a multipart tariff of $15 per connection and a decreasing block volumetric charge for water consumed during the month of June, what is the total charge for each group?

The three blocks are as follows:

- Block 1: $0.06 per cubic meter for first 600 cm of water consumed
- Block 2: $0.04 per cubic meter for amounts between 601 cm and 1100 cm of water consumed
- Block 3: $0.02 per cubic meter for amounts over 1100 cm of water consumed

Part F:

Assuming all 175 customers are charged a multipart tariff of $15 per connection and a uniform seasonal volumetric charge of $0.05 per cubic meter for water consumed during the summer and $0.03 per cubic meter for water consumed during the winter, what is the total charge for each group? We will assume the amount of water consumed by all groups is constant throughout the year.

Part G:

Assuming all 175 customers are charged a multipart tariff of $15 per connection and a uniform volumetric charge with variations in the per unit charge depending on the zone in which consumers reside. What is the total charge for each group based on the zone structure below?

- Zone 1: \$0.04 per cubic meter of water consumed
- Zone 2: \$0.07 per cubic meter of water consumed

Answer #1:

D) Marginal Cost

Answer #2:

Financing Option	Annual Operating Cost (\$)	Annual Project Cost (\$)	Total Annual Cost (\$)	Full Cost Recovery Tariff (\$ per HH)	Percent Change in Tariff
100% Grant	*130,000*	*0*	*130,000*	*16.25*	*8.3*
65% Grant, 45% Loan (seven year, 15%)	*130,000*	*1,317.86*	*131,317.86*	*16.42*	*9.5*
65% Grant, 45% Loan (14 year, 8%)	*130,000*	*681.43*	*130,681.43*	*16.34*	*8.9*
50% Grant, 50% Loan (seven year, 15%)	*130,000*	*1,464.29*	*131,464.29*	*16.43*	*9.5*
50% Grant, 50% Loan (14 year, 8%)	*130,000*	*757.14*	*130,757.14*	*16.35*	*9*
45% Grant, 65% Loan (seven year, 15%)	*130,000*	*1,903.57*	*131,903.57*	*16.49*	*9.9*
45% Grant, 65% Loan (14 year, 8%)	*130,000*	*984.29*	*130,984.29*	*16.37*	*9.1*
100% Loan (seven year, 15%)	*130,000*	*2,928.57*	*132,928.57*	*16.62*	*10.8*
100% Loan (14 year, 8%)	*130,000*	*1,514.29*	*131,514.29*	*16.44*	*9.6*

Answer #3:

Part A:

Group A: \$30
Group B: \$30
Group C: \$30

Part B:

Group A: $600 * \$0.04 = \$24$
Group B: $900 * \$0.04 = \$36$
Group C: $1400 * \$0.04 = \$56$

Part C:

Group A: $15 + (600 * $0.04) = $39
Group B: $15 + (900 * $0.04) = $51
Group C: $15 + (1400 * $0.04) = $71

Part D:

Group A: $15 + (600 * $0.02) = $27
Group B: $15 + [(600 * $0.02) + (300 * $0.04)] = $39
Group C: $15 + [(600 * $0.02) + (500 * $0.04) + (300 * $0.06)] = $65

Part E:

Group A: $15 + (600 * $0.06) = $51
Group B: $15 + [(600 * $0.06) + (300*$0.04)] = $63
Group C: $15 + [(600 * $0.06) + (500*$0.04) + (300 * $0.02)] = $77

Part F:

Summer = Group A: $15 + (600 * $0.05) = $45
Group B: $15 + (900 * $0.05) = $60
Group C: $15 + (1400 * $0.05) = $85
Winter = Group A: $15 + (600 * $0.03) = $33
Group B: $15 + (900 * $0.03) = $42
Group C: $15 + (1400 * $0.03) = $57

Part G:

Zone1 = Group A: $15 + (600 * $0.04) = $39
Group B: $15 + (900 * $0.04) = $51
Group C: $15 + (1400 * $0.04) = $71
Zone 2 = Group A: $15 + (600 * $0.07) = $57
Group B: $15 + (900 * $0.07) = $78
Group C: $15 + (1400 * $0.07) = $113

11 Subsidies

INTRODUCTION

This is the eleventh in a series of eleven teaching units on the financing of environmental projects.

The first six units in this series constitute the first six steps in developing the financial information necessary for undertaking projects where income is associated with the delivery of certain utility services, whether the utility is publicly or privately owned. This includes projects in the following utility sectors: water, wastewater, solid waste, and certain types of energy efficiency projects. Those six chapters are written from the perspective of a utility executive who needs a project to improve his system and is interested in learning the financial techniques necessary to obtain the money to carry out the project.

After introducing the concept of project finance (Chapter 1), the first step was the presentation of methods for measuring income (Chapter 2 – Measuring Income). Methods for maximizing income were presented in the succeeding chapter, Maximizing Cash Available for Debt Service). It was then emphasized that the maximization of income was of paramount importance because if income exceeded operating expenses, needed projects could be undertaken to improve utility systems by having the utility incur debt. If they had excess income, they would be free of a pernicious reliance on grants. Excess income can be used to pay annual debt service. The next chapter, therefore, was devoted to the explanation of the basic concepts of loans and debt (Chapter 4 – Loan Basics).

Once the basic concepts of borrowing are understood, the next step is to be able to evaluate various types of loans in order to determine which available loan is best for a particular utility. Techniques for evaluating loans, therefore, were presented in the following chapter, Project Valuation.

The next teaching unit used all of the information presented in the preceding five to illustrate which projects were financially feasible for a utility to undertake. It demonstrated both how the largest possible project could be undertaken at the lowest possible cost and how a utility's excess operating income could be used to improve the system (Chapter 6 – Financial Feasibility).

In Chapter 7 – Alternate Finance Sources the perspective shifts. Financial concepts were no longer presented from the utility executive's perspective. Rather, in this chapter, financial concepts were presented from the perspective of a government official whose responsibility it is to manage or administer a program to provide funds for municipal utility services. In other words, in the first six chapters, project finance was discussed from the perspective of one needing funds (for a particular project). In the seventh chapter, the concepts of project finance were discussed from the perspective of a government official who has funds, or whose job it is to seek funds for utility projects in his country. It is this official's responsibility to provide the best quality of utility services to the largest amount of people by providing funds to utility operators for projects, which improve that utility's services to its users.

In the eighth chapter, Sources of Funds, the same perspective was maintained as in the previous chapter. In other words, it was written for a government official whose responsibility is to manage a program to finance projects for utilities. In that chapter, sources of external funding for such programs were examined and compared with a view to the most likely and most reliable sources of funding for utility projects.

In Chapter 9 – Cost/Benefit Analyses, government officials were presented with a basic rubric of how to optimize often meager funds available for environmental projects. Matrices were devised to prioritize the "projects which bring the greatest amount of public health benefits to the largest number of people for the least amount of grant funds".

In Chapter 10 – Tariff Design, we returned to addressing the needs of local government officials and utility managers. In Chapter 3 – Maximizing Cash Available for Debt Service, the need for creating fair, volumetric tariffs that provided enough income to cover the cost of operating and maintaining the utility system was discussed. In this chapter, the methodology of creating such tariffs was presented. There was also a discussion of how such tariffs can be regulated by local governments to assure users of the basic fairness of the tariff design.

In Chapter 11 – Subsidies, another matter that was covered briefly in Chapter 3 – Maximizing Cash Available for Debt Service will be discussed: the role of subsidies in maximizing utility revenues.

The conceptual basis for any subsidy is that the goods or services subsidized would be otherwise unaffordable to consumers. Subsidies are often set and left in place for years, regardless of the intervening growth in income of those using the service. This is especially so in countries approaching market economies. Often, subsidies are left in place long after there are substantial rises in income for the general population.

That being said, there is always a segment of the population whose income does not keep pace with the growing economy. There will always be people who need a subsidy.

The thesis of this teaching unit is that only those truly in need should be subsidized. Those who can afford the service should pay for it. This unit will discuss how to design subsidies that target only the truly needy.

The beginning point for any financial discussion of subsidies is twofold. First, a utility must absolutely minimize its cash expenses – so that it only needs to collect a minimum amount from its users. The second point is that, once the minimum cash requirements are known, a tariff must be designed (See preceding chapter) to bring in the income necessary to pay all expenses, plus any additional funds needed for debt service.

Once a "full cost recovery tariff" has been designed, the question of how to deal with the inability of certain truly needy users to pay their share of the tariff can be addressed.

CHAPTER CONTENTS

Water Subsidy Design

Four important criteria that need to be taken into account when incorporating subsidies into the design of water tariff structures:

1. Genuine need
2. Accurate targeting
3. Low administrative costs
4. No perverse incentives

3

Water Subsidy Design

1. Genuine need

- Question from the outset whether any particular group of water consumers really merits a subsidy, and if so, why
- Study what percentage of household income is being spent on water and/or examine what people are able or willing to pay for improved water services

4

Water Subsidy Design

2. Accurate targeting

- Even if a genuinely needy group of customers have been identified at an aggregate level, it is not always straightforward to identify the individuals who belong to this target group
- Targeting variables should be employed in subsidy schemes to identify households who are eligible to benefit, for example, this can be the level of water use (as in IBT)

5

Water Subsidy Design

2. Accurate targeting (cont.)

- If targeting variables are not well chosen, subsidy funds end up being wasted on households who meet eligibility criteria but who are not genuinely needy
- Basic problem with finding good targeting variables is that once built, water systems remain fixed in time and space while poor households move and change (they migrate, some become wealthier; others can be evicted from rental housing)

6

Water Subsidy Design

3. Low administrative costs

- While it is important to screen customers carefully for subsidy eligibility, the screening process can itself be quite costly in administrative terms
- Balance the need for targeting accuracy against the associated administrative costs

Water Subsidy Design

4. No perverse incentives

Using water tariffs as a means of redistributing income between different customer groups can lead to serious conflicts with the efficiency objective, because it often introduces perverse incentives for households and industrial users to use or not use water.

8

This chapter will discuss the following four topics which are the major factors in formulating policy for subsidies for water and wastewater services:

- The Need for Subsidies
- Delivering the Subsidy
- Targeting Mechanisms
- Affordability Measures

THE NEED FOR SUBSIDIES

In every water system, there will be some families that cannot afford to pay the market price for water services; however, since water is a necessity for life it is the responsibility of the community or government to make certain that such families are given adequate access to water services. There are many affordability mechanisms available to decision makers, who will have to choose the mechanism that is most appropriate for their water utility system. The primary considerations are:

- *How much can the local population afford to pay?* In the US, this is measured using the median household income[1].
- *How much is too much to pay for water?* As a society, what percentage of income is the most that an average family should pay? In the US, the Environmental Protection Agency has set this standard at 2.5% of the median household income[2].

Once a standard of affordability has been decided, then the next step is to consider the following:

- *Identifying the targets for the subsidy*: selecting standards of income to decide who is eligible to receive the subsidy (e.g. any household with income below a certain amount); this can be those households whose income falls in the lowest 20% out of the entire community.
- *Accurate targeting of the subsidy*: once an income standard is set, how to direct the subsidy to the eligible households.
- *Keeping administrative costs low*: selecting the least expensive option for the utility for delivering the subsidy to eligible households.
- *Avoiding perverse incentives*: subsidy design must consider the potential effects that the subsidy may have on the behavior of the recipients; no subsidy should give recipients an incentive to use water wastefully, or to sell it.

1 Median Household Income: if the household income levels of all of the constituents of the water utility system were listed from highest to lowest (or lowest to highest) the median household income would be value (i.e. income) in the center of this range.

2 The 2.5% standard for water rate affordability has been criticized for not being an accurate measure of affordability since it only captures the affordability of water for the middle-income groups; the affordability of rates for low income groups (for whom affordability programs are targeted) is not captured in this measure.

DELIVERING THE SUBSIDY

There are three general ways of delivering the low income subsidy: supply side, demand side, and cross-subsidy.

- *Supply Side*: when the government or other external entity makes resources available to cover the deficit between the costs of service provision and the level of the water bill; resources are transferred directly to the utility and delivered to eligible customers through the tariff structure.

Water Subsidy Design

Direct Subsidies

- The government or some other external entity makes resources available to cover the deficit between the cost of service provision and the level of the water bill
- These resources can be transferred directly to the utility and delivered to customers through the tariff structure (known as *supply-side subsidies*)

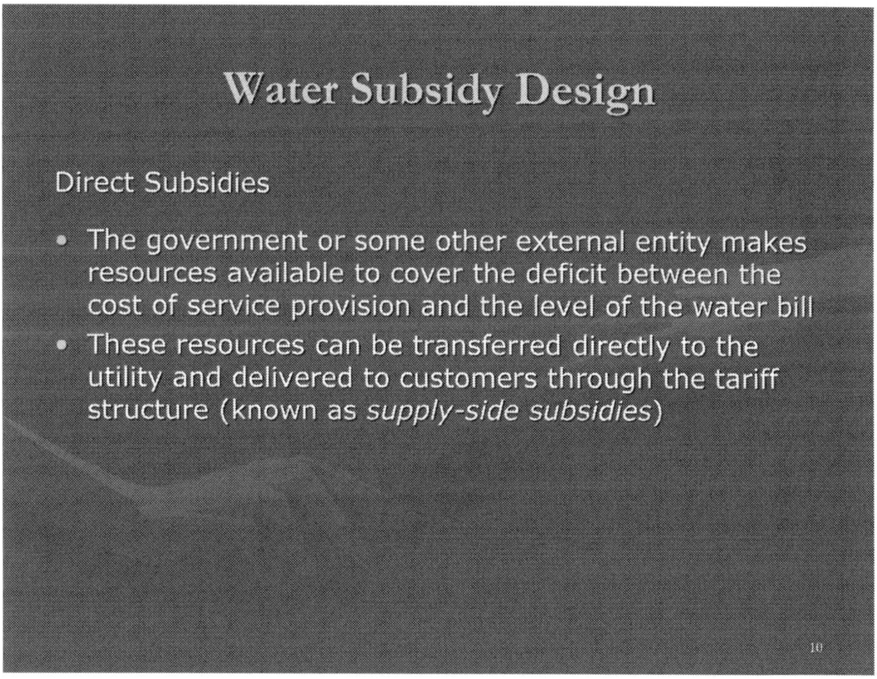

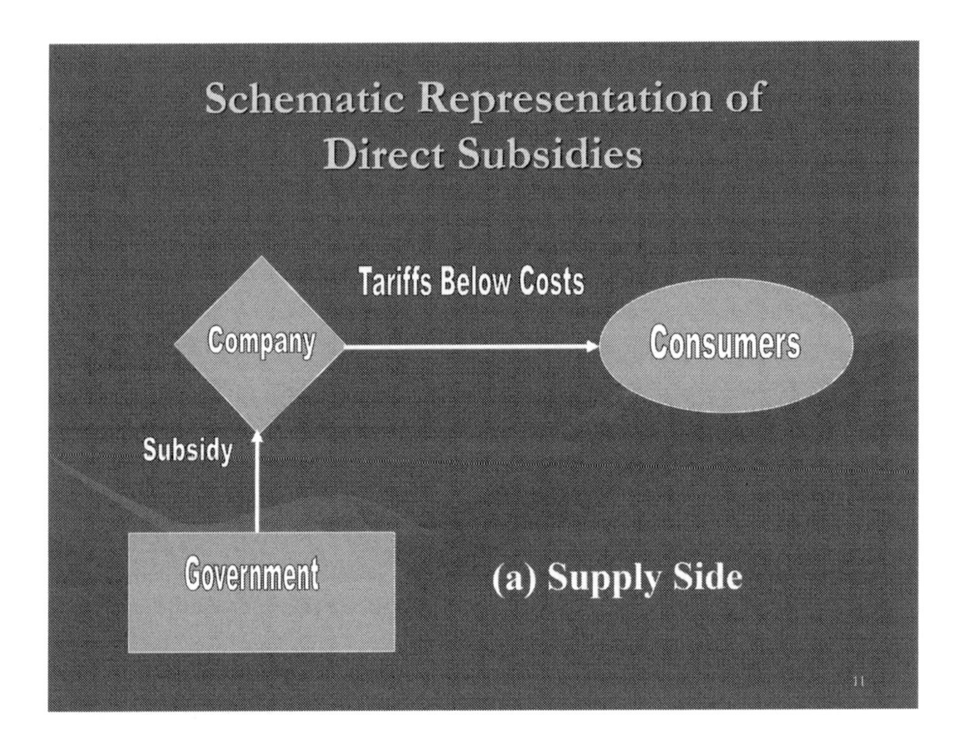

Water Subsidy Design

Supply vs. Demand Side Subsidies

Supply-side subsidies:

- The traditional approach used to subsidize water utilities
- Experience shows that they are problematic
- The presence of major state transfers makes utility managers less concerned about controlling the cost and generates inefficiency
- Tend to lower the general tariff level for all customers often resulting in failure to reach the poor in the way that was anticipated

14

- *Demand Side*: resources are given directly to the customers who are deemed to be eligible; these are usually done outside of the tariff structure.

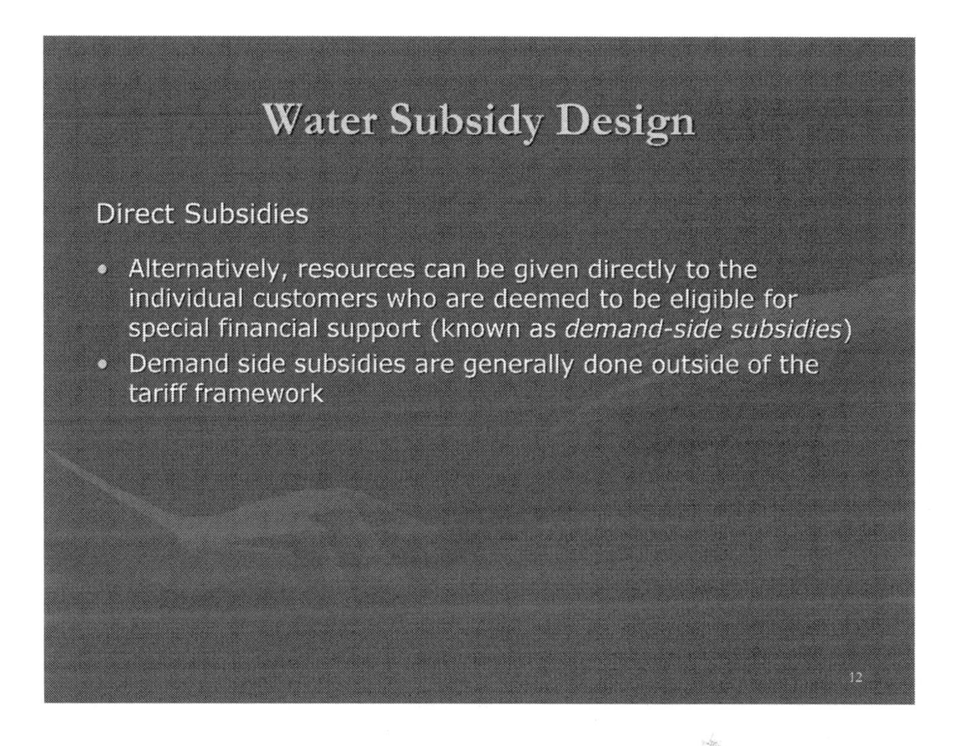

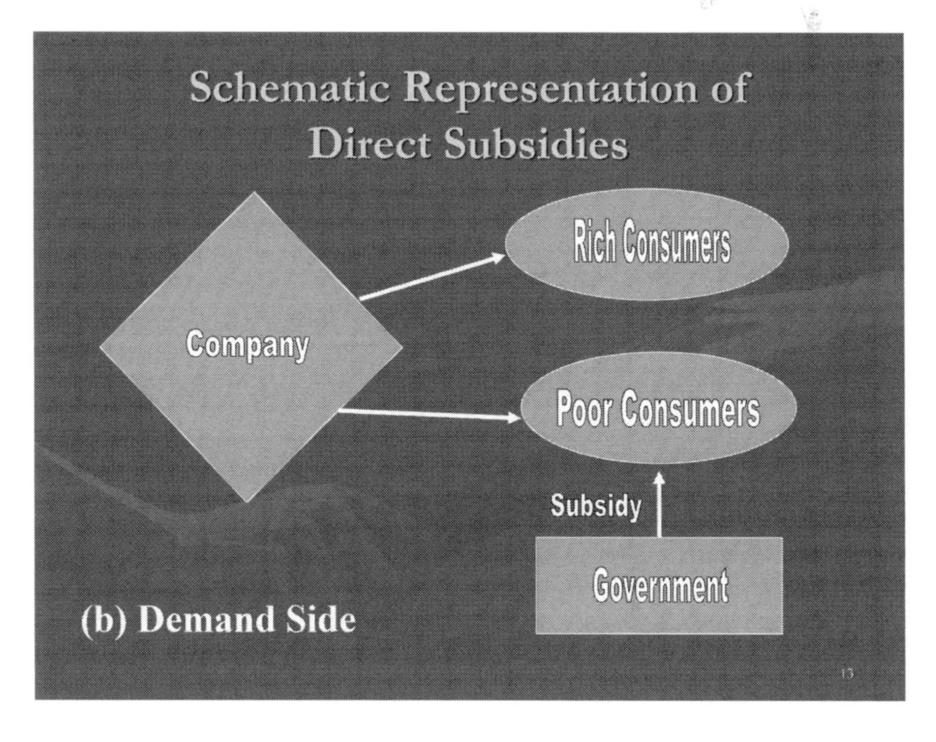

Water Subsidy Design

Supply vs. Demand Side Subsidies

Demand-side subsidies:

If subsidies are necessary, there is a growing preference for demand-side subsidies that go directly towards covering the water bill of the poor household rather than general budget support for the utility.

15

- *Cross Subsidies*: some groups of customers are charged more than the true cost of service provision and this surplus is used to cover the deficit on another set of customers, who pay less than the true cost of service provision.

Water Subsidy Design

Cross-Subsidies

- If government finance is not an option, cross-subsidies can be used whereby some groups of customers are charged more than the true cost of service provision, and this surplus is used to cover the deficit on another set of customers, who pay less than the true cost of provision
- In practice, cross-subsidies and direct subsides are not mutually exclusive, and a large number of public utilities use both simultaneously

16

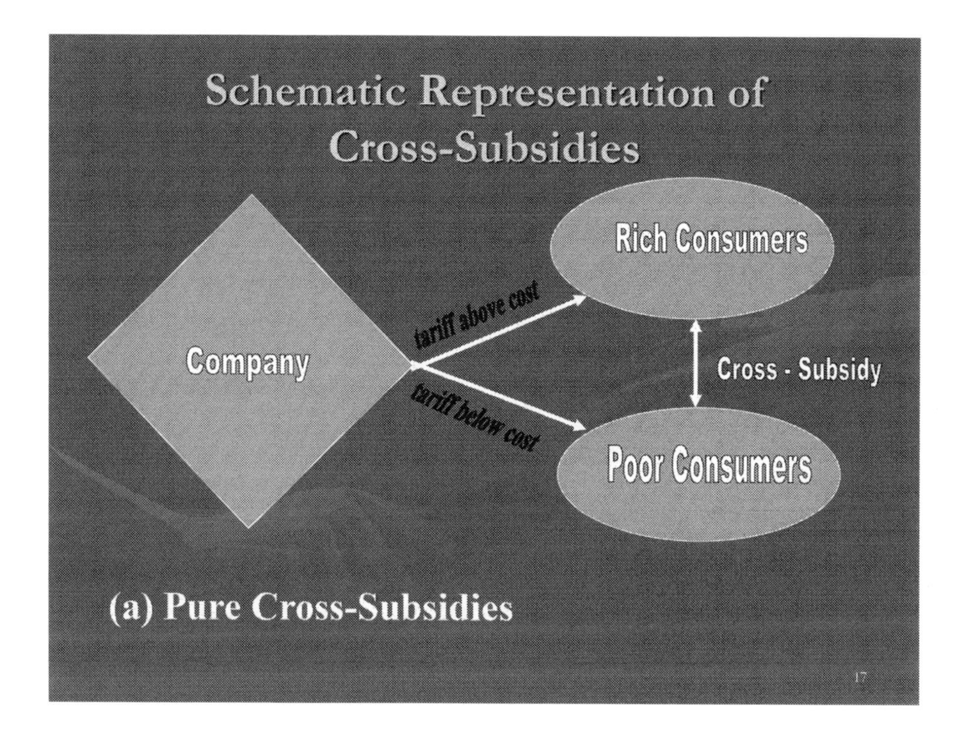

Supply-side subsidies have been the traditional approach used to subsidize water utilities. Experience shows that they are not as effective as demand-based subsidies. With supply-side subsidies, the presence of major state transfers makes utility managers less concerned about controlling the cost and generates inefficiency. Supply-subsidies tend to lower the general tariff level for all customers and hence often fail to reach the poor in the way that was anticipated. If subsidies are necessary, there is a growing preference for demand-side subsidies that go directly towards covering the water bill of the poor household rather than covering the general budget support for the utility.

In practice, cross-subsidies and direct subsides are not mutually exclusive, and a large number of public utilities, use both simultaneously.

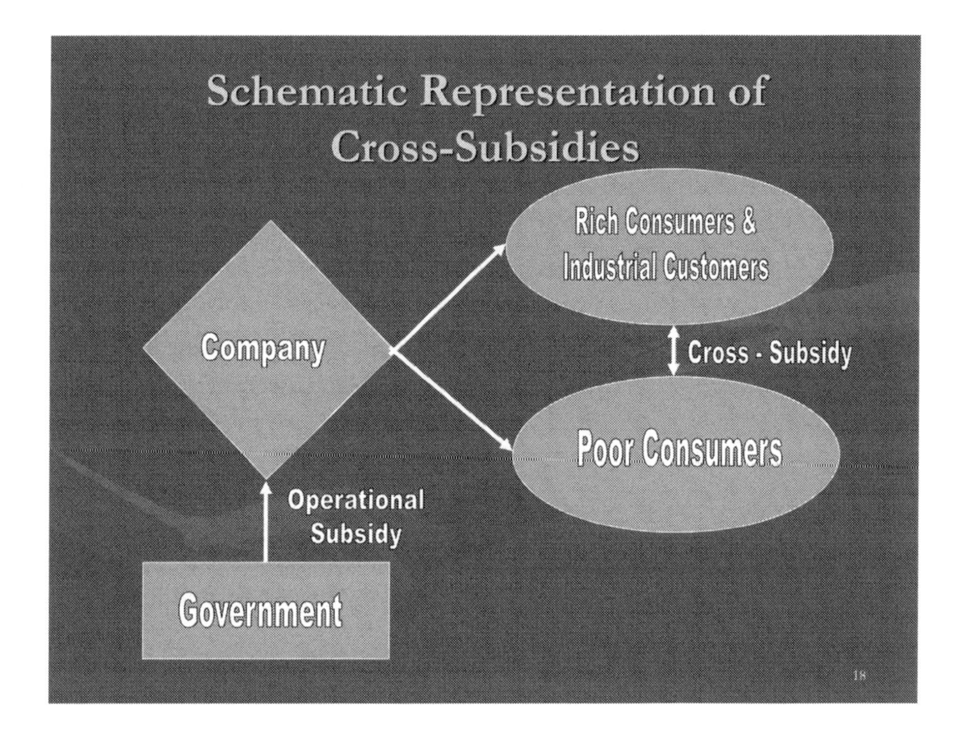

TARGETING MECHANISMS

There are three ways of identifying beneficiaries in order to target subsidies:

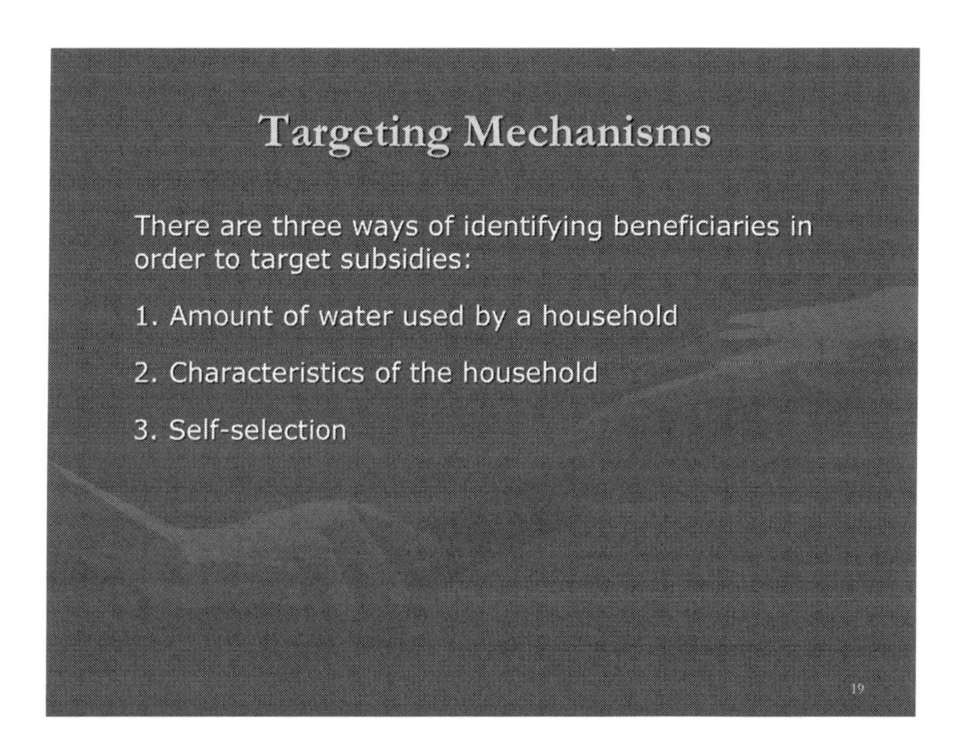

- *Amount of water used by a household*: used with the increasing block tariff (See Chapter 10 – Tariff Design) or with a lifeline rate (described below) so that a small amount of water is provided below cost to all households.

Targeting Mechanisms

1. Amount of water used by a household

- IBTs are often proposed because they are expected to provide a low cost lifeline amount of water to poor households
- However, IBTs provide this subsidy to all connections, regardless of household income level
- Furthermore, household water use is not a good indicator of poverty, because poor households may actually consume relatively large amounts of water, for example, if they have large families or if multiple poor families share a single connection

20

- *Characteristics of the household*: indicators such as geographical location, type of dwelling, income level, or household eligibility for other government assistance programs.

Targeting Mechanisms

2. Characteristics of the household

For example:

- Geographical location
- Type of dwelling
- Income level
- Household eligibility for other governmental assistance programs

21

Targeting Mechanisms

2. Characteristics of the household (cont.)

- Geographical criteria only works in cities that have well defined localized areas of poverty
- However, even then, it is often the case that a large proportion of the poor do not necessarily live in slums but are scattered
- Furthermore, value of subsidies can be capitalized into property values and rents, and thus captured by landlords, not poor tenants

22

- *Self-selection*: the utility provides two distinct levels of service, a high quality service at full cost, and low quality service at subsidized cost given the assumption that only a genuinely poor person would choose the low-quality subsidized service.

Targeting Mechanisms

3. Self-selection

- Under this approach, the utility provides two distinct levels of service:
  - High quality service at full cost
  - Low quality service at subsidized cost
- The idea is that only a genuinely poor person will choose the low-quality subsidized service, because anyone who could afford it would prefer the high quality service
- An example of the self-selection approach would be to subsidize very narrow pipe diameter connections that only provide a limited flow of water into the household

23

The amount of water used by a household is a poor indicator of poverty because poor households may actually consume relatively large amounts of water due to large families, many families sharing one pipe, and leaky appliances in need of repair. Geographical criteria only work in cities that have well defined, localized areas of poverty; however, it is often the case that a large proportion of the poor do not necessarily live in slums but are scattered. Furthermore, the value of subsidies can be capitalized into property values and rents, and thus captured by landlords, not poor tenants.

AFFORDABILITY MEASURES

Given the important considerations above, local government officials and utility managers must choose the type of affordability measure that will accurately target needy households with a subsidy that will help these households use a reasonable amount of water from the utility system. There are four methods available for delivering a subsidy to poor water customers: lifeline rates, refunds or discounts, community social service funds, and national funds for low income water bill assistance.

LIFELINE RATES FOR LOW INCOME CUSTOMERS

Lifeline rates are, in general, a small amount of water (i.e. an amount for minimum essential use) at a rate below the cost of service, while pricing higher-consumption blocks at higher rates. There are two basic types of lifeline rates: those that are applied to all customers for that small block of water used, and those that are only charged to customers who have proven their low income status. Implementing a lifeline rate requires alterations in the pricing structure in order to separate blocks of water use for varying prices, in which case an increasing block rate must be created, and/or to create separate classes of residential customers: those that are eligible for the lifeline vs. those that are not.

If the lifeline rate is limited to low income customers, then it will achieve target efficiency (i.e. help only those who need it); however, if the lifeline is available to all customers then the subsidy is gained by low use customers, who are not necessarily low income – and as such is not fair because this policy gives the benefit to the wrong group of beneficiaries (high-income, low-use individuals).

The lifeline rates are often criticized when they are targeted to low income customers because other customers are forced to pay for the subsidy in the form of increased costs for water consumption. When deciding to implement a lifeline rate, utility managers are advised to seek input from customers to determine their willingness to provide such a subsidy; when subsidies are fair and well-targeted, many people should be willing to accept a slightly larger tariff in order to provide water services to all members of the society.

REFUNDS OR DISCOUNTS FOR LOW INCOME CUSTOMERS

As an alternative to adjusting the rate schedule in order to provide a lifeline rate, many utilities prefer to direct subsidies to eligible low income households by way of a refund or a discount on the households' water bills. The amount of discount offered depends on the full cost recovery tariff required for the utility's revenue to cover its costs and the generally accepted poverty level within the community. If the full cost recovery tariff is 5% of the median household income, a low income person may be paying 7% or more of their income toward the water bill (depending on the tariff level and the income defined as poverty status). Assuming that the local government and community members have decided on a maximum acceptable limit that a household should have to pay for water with respect to its percent of income, then the amount of the discount should equal the percentage difference between the maximum limit and the actual percent of household income that a low income household would pay toward their water bill without the subsidy. For example, in the case where the tariff is equal to 5% of the median household income and the low income households are actually paying 7% of their income toward their water bill, the discount required to enable low income households to pay only 5% of their income towards water is 40%.

The important question that must be addressed once the amount of discount is determined is who will pay for the discount? In order for the utility to continue to cover their costs of water delivery, they will have to charge all other customers higher rates in order to make up the lost revenue from the low-income discount (subsidy). Some local governments are willing to pay for the subsidy from their tax collections, as a form of social service for needy households. The benefit to the utility of having discounts or lower rates for low income customers is the increased likelihood of collecting payment from these customers; the subsidy makes it possible for these customers to pay some of their bill and they will be more likely to pay their bills regularly under this type of program. Just as in the case of the lifeline rate, the utility must seek outside assistance and input when deciding how much of a subsidy to provide and to whom they should provide it.

Community Social Service Funds

The previous two methods have required that a low income customer program be established within the utility. However, the same goal can be otherwise accomplished. A community social service fund can provide financial assistance to low income households outside of the realm of the utility. The utility can charge its low income customers the same amount as its other customers; but the low income customers can then seek bill-paying assistance from the local social service fund. The utility's role in this program is to refer low income customers to the community service fund. This usually occurs when a customer contacts the utility due to their inability to pay the water bill.

There is a downside, however, to the use of community social service funds. Such funds are usually set up by a few concerned community members that are interested in assisting needy households and are dependent on voluntary contributions from members of society. The funds available in these funds are therefore not reliable and are not going to help improve the long-term affordability of water rates for low income customers. In an impoverished community, there is not likely to be enough assistance money available to meet all of the needs of all needy households.

National Funds for Low Income Water Bill Assistance

A national program to provide bill-paying assistance to low income families can potentially work in a similar way to the local community social service funds. A national level fund can help to make certain that low income customers in all communities have access to bill paying assistance. Such a program would be particularly helpful in rural communities where many of the residents are eligible for low income status. Utilities would benefit from the program because it would allow more customers to pay their water bills, which contributes to the ability of the utility to recover its costs of providing the service.

Conclusions

- The design of tariff structures is challenging because there are a number of conflicting objectives involved
- A tariff design that contributes to the achievement of one objective may be detrimental to the achievement of another
- Policy makers need to decide which objectives are the highest priority, and where possible, use more than one instrument
- Fixed charges, widespread in some countries, is the most problematic policy because it generally fails to achieve at least three of the four key policy objectives

24

Conclusions

- IBT structures have often failed to simultaneously meet all of the different objectives of the tariff design
- This is partly due to a poor design of block structures, but also due to the fact that:
  - Low-income households are not necessarily small water consumers
  - Sometimes several poor households share a single connection
- Uniform volumetric rates, whether as a single or two part tariff structure, do comparatively well in meeting the different policy objectives
- In many countries, comprehensive water and sanitation reform will likely require a new institutional framework for the delivery of water services, different from the one that currently exists
- Without sound tariff and subsidy policy, institutional reforms cannot work

25

Questions and Answers

Question #1:

Which of the following is not a major factor in formulating policy for subsidies for water and wastewater services?

A) Delivering the Subsidy
B) Targeting Mechanisms
C) Self-selection
D) The Need for Subsidies
E) Affordability Measures

Question #2:

What are the three general ways of delivering the low income subsidy?

A) Supply side, demand side, and tax exemptions
B) Supply side, demand side, and cross subsidies
C) Cross subsidies, tax exemptions, and grants
D) Supply side, loans, and direct
E) Direct, tax exemptions, and grants

Question #3:

There are three ways of identifying beneficiaries in order to target subsidies:

1. Characteristics of the household
2. Amount of water used by a household
3. Cost/Benefit Analysis

Which of the following statement(s) is/are true?

A) 1 only
B) 1 and 3
C) 2 and 3
D) 1 and 2
E) 1, 2, and 3

Question #4:

The amount of water used by a household is a poor indicator of poverty because poor households may actually consume relatively large amounts of water due to:

1. Large families
2. Many families sharing one pipe
3. Leaky appliances in need of repair

Which statement(s) is/are true?

A) 2 only
B) 1 and 2
C) 1 and 3
D) 1, 2, and 3
E) 3 only

Answer #1:

C) Self-selection.

Answer #2:

B) Supply side, demand side, and cross subsidies.

Answer #3:

D) 1 and 2.

Answer #4:

D) 1, 2, and 3.

Index